3D Local Structure and Functionality Design of Materials

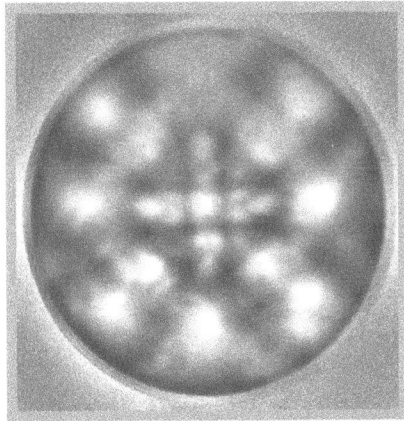

3D Local Structure and Functionality Design of Materials

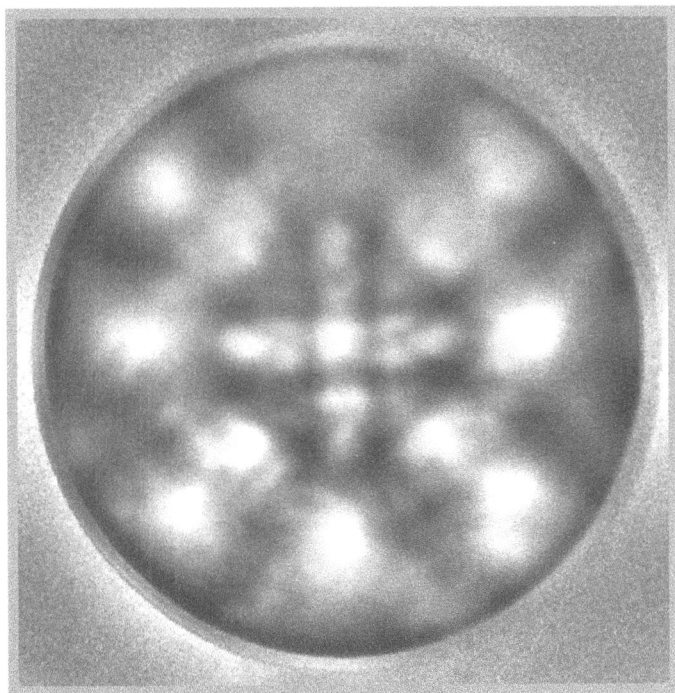

Editors

Hiroshi Daimon
Nara Institute of Science and Technology
Japan

Yuji C. Sasaki
The University of Tokyo
Japan

Ⓜ **MARUZEN** PUBLISHING

🌐 World Scientific

Published by

World Scientific Publishing Co. Pte. Ltd.
5 Toh Tuck Link, Singapore 596224
USA office: 27 Warren Street, Suite 401-402, Hackensack, NJ 07601
UK office: 57 Shelton Street, Covent Garden, London WC2H 9HE

and

Maruzen Publishing Co., Ltd.
2-17, Kanda, Jimbo-cho, Chiyoda-ku,
Tokyo 101-0051, Japan

British Library Cataloguing-in-Publication Data
A catalogue record for this book is available from the British Library.

3D LOCAL STRUCTURE AND FUNCTIONALITY DESIGN OF MATERIALS

ISBN 978-981-3273-66-5

For any available supplementary material, please visit
https://www.worldscientific.com/worldscibooks/10.1142/11083#t=suppl

Desk Editor: Rhaimie Wahap

Typeset by Stallion Press
Email: enquiries@stallionpress.com

PREFACE

This book is the first textbook written in English explaining the new science of *3D local functional-structure science*. This new science has become possible through the application of several *atomic-resolution holographies* to functional materials science. *Atomic-resolution holographies* are new holographies which can directly visualize the 3D local atomic structure around a specific, small amount of active-site atoms. Atomic-level analyses of the mechanisms of functional materials in a wide range of fields such as physics, chemistry, devices, bio-materials, becomes possible by analyzing the atomic structure around active-site atoms governing the materials' function. Functional materials have currently been developed through hypothesis, but from now on, will be developed through experimental evidence. This book describes in easy terms the principle, experimental method, analytical method, and recent achievements of these atomic-resolution holographies.

We are surrounded by a multitude of functional materials, such as the many elements in a mobile phone. Most recent functional materials are typically not pure materials but often have various elements added to create high functionality. For example, in the semiconductor industry, a small amount (about 0.1% or less) of dopant (impurity) is added to a pure Si crystal to activate it as a p- or n-type semiconductor. However, there is no experimental evidence as to whether the dopant atoms exist. Since such a local structure does not have periodicity, X-ray diffraction or electron diffraction — which are standard atomic structural analysis methods — cannot be applied, and the development of functional materials has been carried out in the absence of definite experimental evidence. This kind of phenomenon (i.e., the inability for atomic structure of local active sites to be elucidated) exists widely across various fields; such as catalysts in the chemical industry, proteins in the biological field, etc. If we can clarify the local atomic arrangement around such active-site atoms, major problems in modern society can be solved, from information, to energy, and bio-technologies.

Atomic-resolution holographies utilize quanta, such as electrons, photons (X-rays), or neutrons, to make holograms from specific atoms. These holograms record phases as well as intensities, resulting in a big advantage compared with x-ray diffraction method recording only intensities. *Atomic-resolution holographies* started in the 1980's with photoelectron and X-ray fluorescence holography, but the accuracy was not enough to make actual analyses. Many accurate holographies were developed in Japan in the 2000's, and a big project of *3D active-site science* was started in 2014 to develop this new field. The theory and most advanced examples of these new holograpies are explained in each chapter in this book. Hence the reader can understand the individual methods by reading the corresponding chapter. I hope this book will stimulate the interest of many students and young researchers in this new field of science, *3D local functional-structure science*, and hope this field will pervade throughout the world.

On behalf of editors
Hiroshi Daimon
July, 2018

CONTENTS

Chapter 1

THE RISE OF LOCAL FUNCTIONAL-STRUCTURE SCIENCE

Hiroshi Daimon

Nara Institute of Science and Technology

1.1. Active-Site and Local Functional-Structure

When a substance exhibits some function, its *atomic composition* and *atomic arrangement* are important. The importance of the atomic arrangement can be understood from the fact that a diamond and graphite, which are made of the same carbon atoms but differ in structure, have completely different features, such as color, hardness and electrical conductance.

Most recent functional materials are typically not pure materials but often have various elements added to exhibit high functionality. For example, in the semiconductor industry, a small amount (about 0.1% or less) of dopant (impurity) is added to a pure Si crystal to activate it as a p-type or n-type semiconductor. Such impurities are supposed to enter the substitutional site of the matrix atoms or interstitial site. It is said that they become electrically active only when they enter the substitutional site, although this is yet to be confirmed. Since such a local structure does not have periodicity, X-ray diffraction or electron diffraction — which are standard atomic structural analysis methods — cannot be applied, and the development of functional materials has been carried out without definite experimental evidence.

This kind of phenomenon (i.e., the atomic structure of local active sites cannot be elucidated) exists widely across various fields. Many solid catalysts are used in the chemical industry, where most of the active sites of the catalytic reactions are specific metal atoms. In photosynthetic proteins in the biological field, the central Mn atom governs the entire photosynthetic reaction. If we can clarify the local atomic arrangement around such active-site atoms, it solves big problems in modern society, from the information technology industry to energy, by contributing to the high efficiency of many functional materials from semiconductor devices, energy devices, and chemical syntheses, to photosynthesis.

This chapter overviews why these local structures cannot be analyzed by conventional X-ray diffraction or electron diffraction and why it can be clarified by methods introduced in this book. Figure 1.1(a) shows a conceptual drawing of conventional X-ray diffraction and electron diffraction. In X-ray and electron diffraction, a thin and parallel incident beam Ψ^0 is incident on the crystal and a scattered beam Ψ^S, which is Bragg-reflected by various faces in the crystal, is observed. If the sample is not a crystal, where atoms are periodically aligned, Bragg reflection

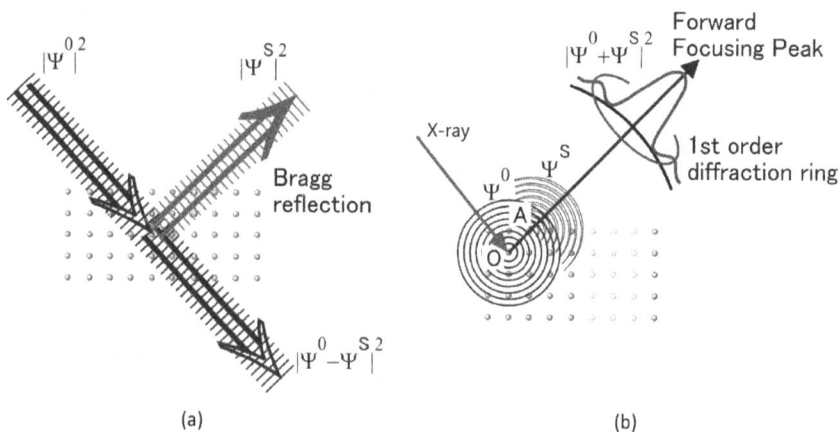

Fig. 1.1   Conceptual diagram of (a) X-ray and electron diffraction (b) photoelectron diffraction and holography.

does not occur and measurement cannot be performed. Even if structural analysis is attempted around the extremely small amount of dopant in Si as mentioned above, it is impossible to measure because the dopants are not arranged periodically.

Photoelectron diffraction[1] shown in Fig. 1.1(b) is a method to elucidate the non-periodic structure around the focused atom at the atomic level. When an X-ray photon is irradiated to an atom, the valence or core electron receives its energy and jumps out into the vacuum. This is subsequently called a photoelectron. When X-rays are irradiated on the sample, photoelectrons are emitted from all the atoms in the crystal, but the energy of the core level varies depending on the element. Therefore, by selecting the photoelectrons' energy, only photoelectrons from the focused element (i.e., the dopant in the case described above) can be detected. The emission angle distribution of the photoelectron intensity is called a photoelectron diffraction pattern. The wave function of photoelectrons Ψ^0 propagates spherically as shown in Fig. 1.1(b). If another atom (A atom in the figure) exists nearby while propagating into the vacuum, Ψ^0 is scattered resulting in the scattered wave Ψ^S. When the kinetic energy of photoelectrons from the core state reaches several hundred eV or more, the de Broglie wavelength of electrons becomes 1 Å or less, so interference between Ψ^0 and Ψ^S is generated to produce a photoelectron diffraction pattern as shown in Fig. 1.1(b). The *photoelectron diffraction* method involves analyzing the local atomic arrangement around the emitter atom by analyzing this pattern. The atomic model structure around the photoelectron emitter atom is judged to be correct if the photoelectron diffraction pattern calculated assuming the atomic model structure matches the measured pattern, hence it is called *indirect* method. The X-ray and electron diffraction are also *indirect* methods to judge the atomic model structure from the degree of agreement between the pattern calculated from the model and the actual measurement.

The *atomic resolution holography* methods, which can inversely convert the measured pattern *directly* to the atomic arrangement, have recently been developed greatly in Japan. It has become possible to discuss the materials property based on the local atomic structure with experimental evidence.

1.2.   Invention of Holography and Its Resolution

Holography was invented by Gabor[2] to increase the resolution of the electron microscope in 1949. Gabor won the Nobel Prize in 1971 with this achievement. The principle of holography is shown in Fig. 1.2. The object is irradiated with a plane wave from the left, and an interference pattern between the scattered wave (object wave) scattered by the object, and the original wave (reference wave) not scattered is recorded as a hologram on the photographic plate. As shown in the figure, if a photographic plate is placed along the ridge of the reference wave, the ridge of the object wave at the hologram is intensified and recorded brightly, and the place of the valley of the object wave is weakened and recorded as a dark contrast. Hence, not only the intensity of the object wave but also the phase is recorded in the interference pattern on the hologram. When reproducing the image, only the reference wave is incident from the left side of the hologram; the intensity and the phase of the object wave are reproduced on the hologram. Hence the observer viewing from the right can see a three dimensional shape of the object. For holography, a beam with good coherency (high

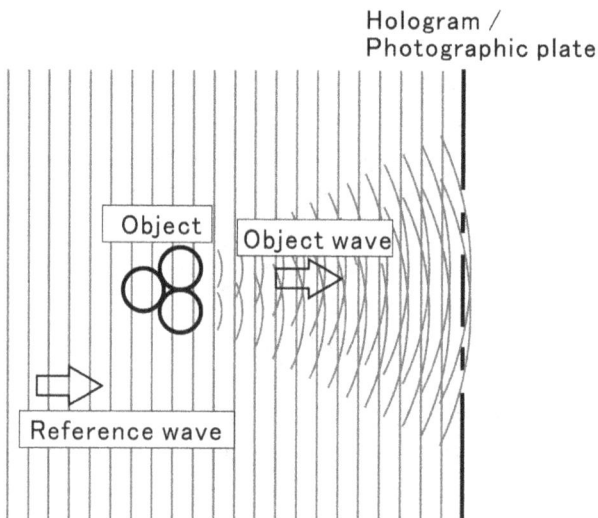

Hologram /
Photographic plate

Object

Object wave

Reference wave

Fig. 1.2   Principle of holography.

parallelism) is necessary. If non-parallel lights are mixed in the reference wave in Fig. 1.2, the interference pattern of the hologram becomes blurred and disappears. In visible light, the laser enables beams with good coherency to be easily obtained, so holograms can be made easily, and are used widely around us, such as on money bills. On the other hand, electron beam holography with atomic resolution has not been realized because it is difficult to make an electron beam with coherence good enough to make a non-blurred interference pattern from an atomic structure.

1.3.   Dawn of Atomic Resolution Holography

In 1986, Szöke[3] proposed that atomic resolution holography could be realized by considering the direct photoelectron and the scattered photoelectrons in Fig. 1.1(b), as the reference and the object waves shown in Fig. 1.2. Since the phase difference between the direct wave and the scattered wave is also recorded in the photoelectron diffraction pattern, it can be considered to be a hologram. This technique is called *photoelectron holography*. In this case, even if you try to see the atoms by irradiating light, the light with the same wavelength as the photoelectron is an X-ray, so it is invisible to the eyes. Even if you can see the X-ray, the atoms are too small to see. Instead, it was shown that atoms can be reproduced by a computer using an equation similar to a Fourier transformation.[4] Since then, a lot of research has been conducted, and efforts to improve the quality of reconstructed images have accumulated. However, the accuracy of the reproduced positions has been as low as 0.5 Å and the number of atoms that can be reproduced was also up to the second nearest atoms. Hence it was hardly used for practical use, and research has halted.

At the same time, *X-ray fluorescence Holography*[5] was also performed to reproduce the arrangement of atoms around a focused atom from the intensity distribution pattern of fluorescent X-rays that emitted from the focused atom like photoelectrons. However, the intensity of fluorescent X-rays was weak, taking about two months to take one hologram, and the accuracy of the reconstructed image was poor, so similarly it did not progress.

1.4.　Renaissance of Holography in Japan

Recent Japanese studies have broken through these stagnations. Matsushita developed a new analysis method in 2005,[6] and the accuracy of the position of the reconstructed image and the number of atoms reproduced in photoelectron holography dramatically increased (see Chapter 2). In X-ray fluorescence holography, Hayashi succeeded to reproduce high-precision images with the aid of strong X-rays from advanced synchrotron radiation facilities and a highly-sensitive detector[7] (see Chapter 2). In addition, *CTR (crystal truncation rod) holography*, which considers the reflected X-ray from a crystal as a reference wave, was also invented by Takahashi in 2001[8] (Chapter 3). In these holographies, since the phase of the scattered wave is recorded in the hologram, there is no *phase problem* which exists in X-ray and electron diffraction. These methods are *direct*, where the atomic arrangement can be directly obtained without assuming a structural model.

Even in these *direct methods*, simple calculations, such as the Fourier transformation, are necessary, and it is impossible to estimate the structure without looking at the calculation results. A technique called *atomic stereophotography*,[9] which can directly take stereophotographs of three-dimensional atomic arrangements without any calculation, was also invented by Daimon in 2001. Thus, recent developments in atomic resolution holography are all done in Japan, and Japan is booming in this field with leading the world.

1.5.　Atomic Stereophotography

The *atomic stereophotography* technique is different from holography, with a different principle that is introduced here. In the photoelectron diffraction pattern of Fig. 1.1(b), a strong forward focusing peak appears in the direction connecting the photoelectron emitter atom and scatterer atoms. This phenomenon occurs because the atomic potential is an attractive potential for electrons, and therefore has an effect like a convex lens. We found that the forward focusing peak rotates about 2° in the direction of rotation of the circularly polarized light when the irradiated X-ray is circularly polarized.[10] Figure 1.3 shows how photoelectrons emitted from atom O are scattered by atom A in a plane perpendicular

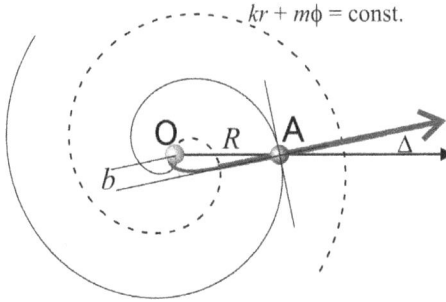

Fig. 1.3   Scattering of photoelectron in a plane perpendicular to the incident circularly polarized photon direction.

to the optical axis of incident circularly polarized light. Solid and dashed lines are equiphase surfaces of the ridge and valley of the photoelectron wave. Since the core photoelectrons excited by a circularly polarized X-ray receives angular momentum ($\hbar$) of circularly polarized X-ray, if the angular momentum held by the core state is $m_0\hbar$, the angular momentum of photoelectron $m\hbar$ is $(m_0 + 1)\hbar$. The equiphase surface (wavefront) of the photoelectron wave with angular momentum is not a concentric sphere but a vortex as shown in Fig. 1.3. Figure 1.3 is calculated assuming that the angular momentum quantum number m of photoelectrons is 4, the kinetic energy is 150 eV, and the distance R between O and A is 2 Å. Since the traveling direction of the wave is perpendicular to the wavefront at the atom A located at the distance R from the emitting atom O, it is oriented along the thick arrow displaced by Δ from the direction of OA. Δ is expressed by the following equation.

$$\Delta = \tan^{-1}\frac{m}{kR\sin^2\theta} \cong \frac{m}{kR\sin^2\theta} \tag{1.1}$$

That is, the direction of the forward focusing peak of the core photoelectron excited by the left- and right-handed circularly polarized X-ray slightly deviates from the direction (θ, ϕ) of the scattering atom resulting in:

$$(\theta, \phi \pm \Delta). \tag{1.2}$$

Since m is determined by the type of the core state, if the Δ is measured, the bonding distance R will be known. Since the direction of the scattering atom is known in the average direction of (1.2), the three-dimensional position (R, θ, ϕ) can be directly measured without detailed calculation.

When looking at three-dimensionally arranged objects with the left and right eyes, the angles seen by the left and right eyes are slightly different depending on the distance from the observer to the object, and the deviation of that angle is called *parallax*. A pair of images for the left and right eyes are stereophotographs, and by seeing them with the left and right eyes, respectively, the original object can be recognized three-dimensionally. The parallax angle $(\pm\Delta)$ is inversely proportional to the distance R from the observer to the object, and is smaller as the distance is longer. Since the relationship between Δ and R is the same as in Eq. (1.1), it was shown that the circularly polarized photoelectron diffraction pattern can be considered as a stereophotograph of atomic arrangement.[9]

Stereophotographs of InP crystals taken in this manner are shown in Figs. 1.4(a) and (b).[12] Looking at the picture Fig. 1.4(a) with the left eye and the Fig. 1.4(b) with the right eye, respectively, the atomic arrangement when viewing the direction of the A atom from the In atom position of O in Fig. 1.4(c) can be recognized stereoscopically. The nearest atom A is seen closer than the second nearest atoms B and C.

Since Δ is small the measurement error of Δ is large, and the accuracy of the interatomic distance R which can be seen (obtained) by this method is not high. However, this method has an advantage that you can see directly without calculation. Even in photoelectron and X-ray fluorescence holography (described later), it is impossible to imagine the structure by merely looking at the data and it is only a matter of believing the calculation result. However, since the approximate position can be known with this method, the reliability of the obtained result is remarkably improved.

1.6. The Rise of *Local Functional-Structure Science* by *3D Active-Site Science*

Based on the atomic resolution holography boom in Japan as described above, systematic research on the local structure around the active-site —

(a)

(b)

C   A   B

Photoelectrons excited by
circularly polarized light

O

(c)

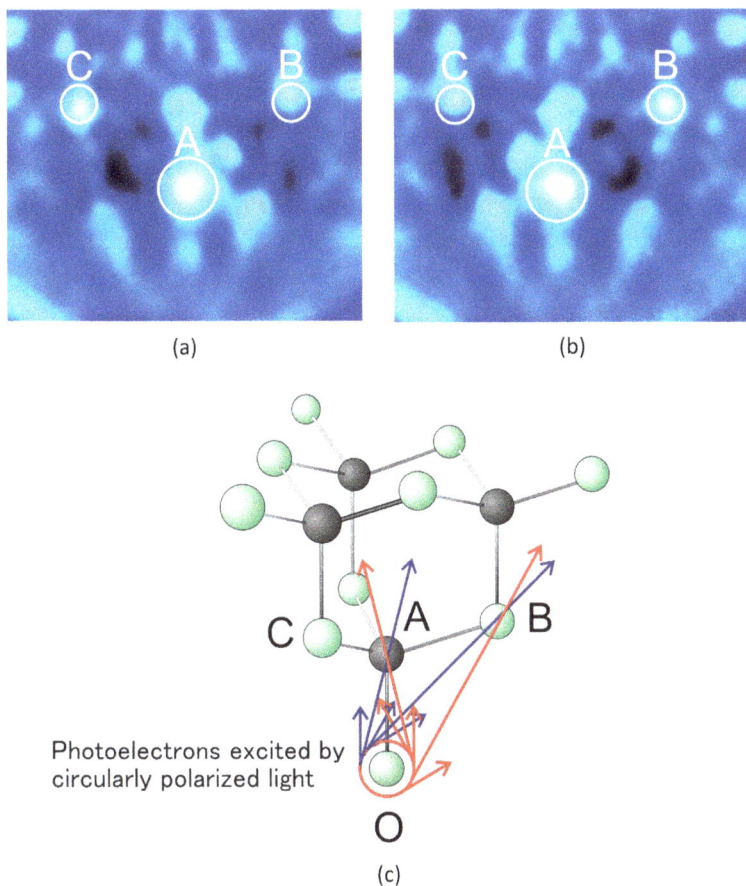

Fig. 1.4   Stereophotograph of InP crystal viewed from In atom O.[12]

to date as a nonperiodic structure has been impossible to analyze — has
started as a project of *3D active-site science*.[11] This project actively collects
the local structural data of a wide range of target substances, understands
the mechanism of various functions theoretically, and through these
strongly promotes the new *local functional-structure science*. Semiconductors and metals have been mainly studied so far, but the target is being
extended to catalysts in chemical reaction systems, organic semiconductors, as well as bio materials such as proteins. The measurement methods
have similarly been extended to neutron holography, electron scattering

holography and others, which enabled to analyze light elements and nanoparticles.

New technologies for energy creation, energy saving, energy storage, etc. are demanded in the fight against global warming caused by massive consumption of fossil fuels, and accidents related to nuclear power. In order to develop new functional materials for these purposes, we are required to solve various problems by a new strategic materials science. Even the development of silicon — the base material of the semiconductor industry — was made by trial and error without knowing the local structure of the dopant. In order to break through the current situation, application of this *atomic resolution holography* is demanded. By systematically solving many problems such as solar cells, catalysts, spintronics materials, light metals, power electronics, secondary batteries, fuel cells, sensors, scintillators for PET (positron emission tomography), etc., we believe that we can contribute to the realization of a *clean, safe and secure longevity society*.

References

[1] C. S. Fadley, *Synchrotron Radiation Research: Advances in Surface Science* (Plenum Press, New York, 1990).
[2] D. Gabor, *Nature* **161**, 777 (1949).
[3] A. Szöke, AIP Conference Proceedings No. 147 (AIP New York 1986).
[4] J. J. Barton, *Phys. Rev. Lett.* **61**, 1356 (1988).
[5] M. Tegze and G. Faigel, *Nature* **380**, 49 (1996).
[6] T. Matsushita, A. Yoshigoe, and A. Agui, *Europhys. Lett.*, **71**, 597 (2005).
[7] K. Hayashi, *et al.*, *Nucl. Instrum. Methods Phys. Res. A* **467/468**, 1241 (2001).
[8] T. Takahashi, K. Sumitani, and S. Kusano, *Surf. Sci.* **493**, 36 (2001).
[9] H. Daimon, *Phys. Rev. Lett.* **86**, 2034 (2001).
[10] H. Daimon, T. Nakatani, S. Imada, S. Suga, Y. Kagoshima, and T. Miyahara, *Jpn. J. Appl. Phys.* **32**, L1480 (1993).
[11] T. Matsumoto, *et al.*, e-J. *Surf. Sci. Nanotech.* **7**, 181 (2009).
[12] http://www.en.3d-activesite.jp/.

Chapter 2

PHOTOELECTRON HOLOGRAPHY AND X-RAY FLUORESCENCE HOLOGRAPHY

PHOTOELECTRON HOLOGRAPHY

Tomohiro Matsushita

Japan Synchrotron Radiation Research Institute

Fumihiko Matsui

Institute for Molecular Science

2.1.1. Introduction

When a solid is irradiated with X-rays, electrons in the core level are excited higher than the vacuum level, and consequently, photoelectrons and Auger electrons are emitted from the surface. Photoelectron spectroscopy is a method of analyzing atomic and electronic structures of solid based on measurement of kinetic energy of these photoelectrons. Atoms are composed of nuclei, electrons at the core level, and valence electrons. Core level electrons strongly bind to nuclei and valence electrons play a role of connecting atoms in molecules and crystals. The density of the state of the valence band can be obtained by measuring photoelectrons excited from the valence band. The chemical composition information of the solid can be elucidated from the core-level photoelectron spectroscopy. Furthermore, by analyzing the angular distribution of the valence photoelectron intensity, the band structure of the crystal surface can be visualized. In the case of core-level excitations, a photoelectron hologram is measured and information on the atomic structure is obtained.

Figure 2.1.1 shows a schematic diagram of the photoelectron excitation process. First, a sample is irradiated with monochromatic light (1), and the core-level photoelectrons are emitted. Photoelectron propagates as spherical wave (2). Part of the photoelectrons are scattered by surrounding atoms and form scattered waves (3). The direct wave and the scattered wave interfere with each other, and a diffraction pattern appears in the photoelectron intensity angular distribution (4). These phenomena are called photoelectron diffraction.

Holography is a technique of generating a three-dimensional image by irradiating a hologram with reproduction reference light. The photoelectron excitation phenomenon can be regarded as a hologram recording process of local atomic arrangement. The first direct spherical wave and the subsequent scattering wave correspond to the reference wave and the object wave, respectively, of optical holography. Hereinafter, the photoelectron diffraction pattern is called a photoelectron hologram. The technique of reconstructing the atomic structure from a photoelectron hologram is called photoelectron holography. The photoelectron

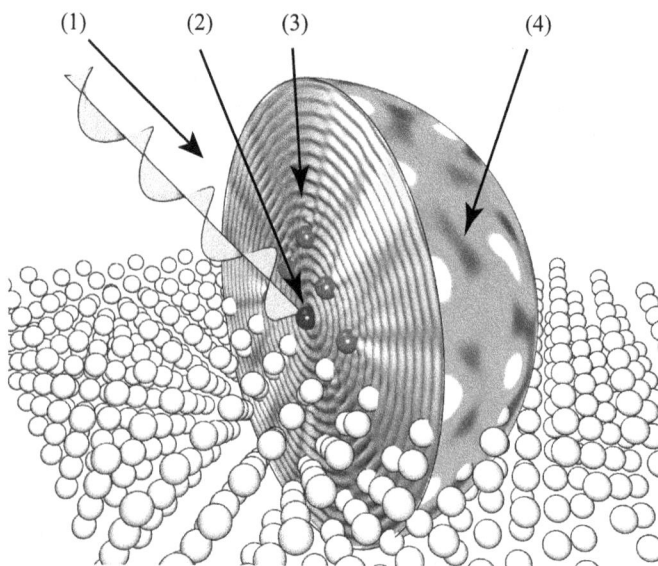

Fig. 2.1.1 Principle of photoelectron holography.

holography is a direct structural analysis method that reconstructs a three-dimensional atomic arrangement without using an initial structural model.

Auger electron emission accompanies photoelectron emission. Auger electron diffraction is a result of Auger electron wave and scattered wave interference similar to the case of photoelectron diffraction. Hereinafter, for convenience, both photoelectrons and Auger electron holograms are called "photoelectron holograms".

Unlike X-ray diffraction (XRD) and electron diffraction based on interference between scattered waves, the local structure around atoms emitting photoelectrons can be reconstructed by photoelectron holography. Each element holds a unique value set of core-level binding energies. Furthermore, the spectral profile of photoelectrons is sensitive to core level chemical shifts that reflect different chemical environments at each atomic site. This made it possible to distinguish atomic sites with different elements and chemical environments. In photoelectron holography, the orientation of the local structure around each emitter atom should be aligned in a specific direction, but the long distance order is not a strictly demanded. Thus, it is possible to analyze local structures around targeted atoms such as impurities in the crystal, embedded interfaces, surface adsorbed species, and thin epitaxial films. Many of the excited electrons are inelastically scattered and lose energy while moving through the solid. The length of the inelastic mean free path (IMFP) is the average distance of electrons traveling in the solid without losing energy during flight. The depth range of the atomic arrangement information contained in the hologram is evaluated by the length of the IMFP. For example, in the case of soft X-ray excitation with a kinetic energy of several hundred eV, most of the photoelectrons come from a surface region within a depth of about 1 nm. In the case of hard X-ray excitation, photoelectrons of several keV have longer IMFP length and bulk-sensitive measurement is realized.

Two-dimensional intensity angular distribution, $I_{E_k}(\theta, \phi)$, of photoelectrons and Auger electrons is measured using angle-resolved electron energy analyzer. E_k is the kinetic energy of electron. Single energy photoelectron holograms are obtained by commercially available laboratory X-ray sources. By using a light source with variable light energy such as synchrotron radiation, a series of photoelectron holograms can be

obtained as a function of kinetic energy. The authors developed a reconstruction calculation method incorporating the scattering process.[1,2] In
this method, since the "phase shift problem" and the "conjugate image
problem" in the conventional method are solved. Atomic arrangement
can be obtained even with a single energy photoelectron hologram.

X-ray fluorescence holography is a measurement technique for elucidating the three-dimensional arrangement of atoms recorded in interference patterns of fluorescent X-rays emitted by local atoms and scattered
by the surrounding atoms. It is possible to reconstruct an atomic image
in the range of 2 to 3 nm around atoms emitting fluorescent X-rays.
This is particularly useful for analyzing dopant atoms, lattice distortion,
and local structures around special clusters. Later in this chapter, we will
explain the basic principle of these methods and introduce examples of
some applications.

2.1.2. Apparatus for Photoelectron Holography

Characteristic X-ray sources are used for photoelectron hologram measurements in laboratories. On the other hand, in the synchrotron
radiation facility, it is possible to obtain monochromatic X-rays of
high quality (intensity, polarization, pulse width, beam size *etc.*) and
variable energy. For research on photoelectron holograms, an electron
energy analyzer having excellent angular resolution is required. Analyzer
has been categorized into two types from the viewpoint of detection
mechanism of two-dimensional angular distribution. They are shown in
Fig. 2.1.2.

2.1.2.1. *One-dimensional analyzer with high energy resolution*

There are two ways to obtain an angular distribution using a one-
dimensional analyzer, *i.e.* an electrostatic concentric hemispherical analyzer (CHA). One way is to move the electron energy analyzer in
a vacuum chamber. However, as the size of the analyzer is limited,
it is difficult to achieve high energy resolution. The other way is to
rotate both the polar angle and the azimuth of the sample. High energy
resolution can be realized. Some types of CHA have two-dimensional
detectors that can measure one-dimensional intensity angle distribution

(a)

(b)

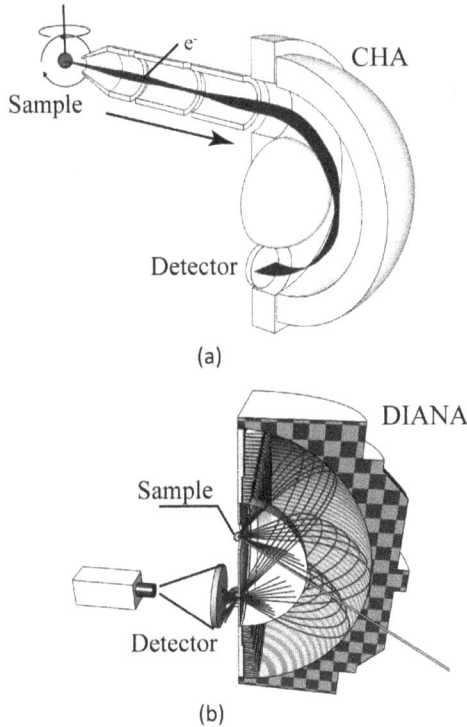

Fig. 2.1.2 (a) The apparatus for the photoelectron hologram measurements using high resolution electron energy analyzer (b) Daimon-type two-dimensional display electron-energy analyzer.

and energy dispersion at a time. In order to obtain a two-dimensional emission angle distribution, measurements are made while scanning the orientation of the sample.

2.1.2.2. *Two-dimensional display type analyzer*

In CHA, entrance and exit apertures are placed between the electrodes of two concentric spheres. By replacing the inner electrode with a mesh structure and shifting the position of the entrance and exit apertures inside the inner spherical mesh, the two-dimensional angular distribution of the photoelectron passing through the exit aperture is projected onto the detector. The so-called two-dimensional display type analyzer has a uniqueness capable of measuring a two-dimensional angular distribution at a time. Typical examples are the Eastman-type ellipsoidal-mirror display analyzer[3] and the Daimon-type spherical-mirror display analyzer.[4] In recent years, analyzers using wide-angle spherical aberration correction with an ellisopidal mesh,[5] a momentum microscope combining two CHAs,[6] a time-of-flight analyzer,[7] a device introducing a deflection mechanism in CHA, *etc.* have appeared. These analyzers are characterized by being able to make maximum use of polarization dependence in photoelectron spectroscopic experiments.

2.1.3. Photoelectron Excitation Process

The excitation light is an electromagnetic wave defined by wave vector $\mathbf{k}_{photon}$, amplitude A_0, and polarization $\mathbf{e}$. The energy of light is proportional to the wave number $k(\text{Å}^{-1}) \simeq 0.5\,E(\text{keV})$. Electrons are excited by this electromagnetic wave and become photoelectrons. Photoelectrons travel through solids and go out into vacuum. By measuring its kinetic energy and emitted direction of photoelectron, we can specify the binding energy and motion of electrons at the initial state.

2.1.3.1. *Conservation of energy*

Total energy is preserved before and after the photoexcitation process. The energy of the final state E_f is given by $E_f = E_i + h\nu$, where E_i and $h\nu$ are the initial state energy and the photon energy, respectively. The kinetic energy E_{kin} of photoelectrons emitted in a vacuum takes a positive value above the vacuum level E_{vac}. On the other hand, the binding energy E_B is defined based on the Fermi level E_F. The difference between the vacuum level and the Fermi level is the work function ϕ.[*1] Therefore, the following relation holds;

$$E_{kin}^{vac} \equiv E_f - E_{vac}, \tag{2.1.1}$$

$$E_B \equiv -(E_i - E_F) = h\nu - E_{kin}^{vac} - \phi. \tag{2.1.2}$$

The kinetic energy distribution of photoelectrons reflects the density of states of electrons. Photoelectron spectroscopy is a technique to clarify

---

[*1] $\phi = E_{vac} - E_F$. Be careful not to confuse with azimuthal angle ϕ.

the energy level of electronic structure based on the law of energy conservation. The zero energy of photoelectrons in a solid is the bottom energy of the valence band. The energy from the Fermi level to the bottom of the valence band is called inner potential V_0. The kinetic energy of photoelectrons when forming a hologram in the crystal is given by

$$E_{\text{kin}}^{\text{holo}} \equiv E_{\text{kin}}^{\text{vac}} + \phi + V_0. \tag{2.1.3}$$

2.1.3.2. *Conservation law of angular momentum*

Before and after the photoexcitation process, total angular momentum (orbital magnetic and spin quantum numbers) is also preserved.

In a dipole approximation, linearly polarized light can be expressed as p_z-orbital-like spherical harmonics $|l = 1, m_l = 0\rangle$, when the z axis is set to the direction of the polarization vector **e**. The angular momentum of light is added to the orbital magnetic quantum number of the electron. Total angular momentum is handed to the photoelectron. The angular momentum is reflected in the photoelectron intensity angular distribution. Figure 2.1.3 shows the atomic orbital in the initial state and the photoelectron intensity angular distribution. From the angular distribution, the atomic orbital in the initial state can be specified.

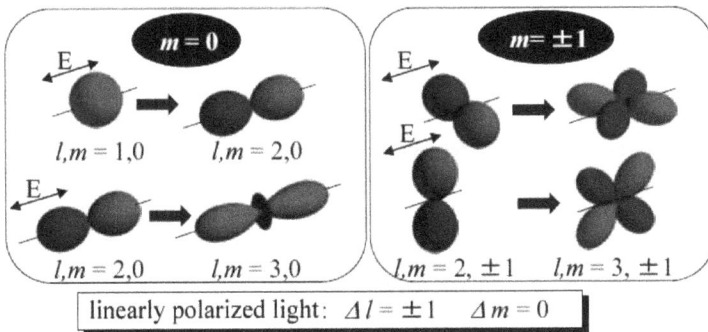

Fig. 2.1.3 Relationship between angular momentum conservation in photoelectron excitation process. The z axis is the direction of the electric field vector. The initial state of $m_l = 0$, s, p_z are excited into p_z, d_{z^2} state, respectively (left panel). The orbit with $m_l = \pm 1$ is excited to photoelectron with intensity angular distribution of $m_l = \pm 1$ (right panel).

For example, the p_z final state is the only destination of transition from $1s$ orbital, while from the $2p_z$ orbital, transition to the d_{z^2} final state as well as to the s final state takes place. Since the photoelectron wave function plays the role of the reference wave of the photoelectron hologram, it is an important factor in the formation of the photoelectron hologram. In the simulation and analysis of the photoelectron hologram, the information on the initial state and polarization of light is important.

2.1.3.3. *Auger electron transition*

When a core hole is formed by photoelectron emission, an electron with a higher energy level falls into the core hole and energy is released. The released energy is transferred to another electron, which is subsequently ejected from the atom. This second ejected electron is called an Auger electron. For core holes deeper than 1 keV, the energy is mainly released as fluorescent X-rays. The Auger electron and fluorescent X ray are not suitable for multi energy hologram measurement because the kinetic energy is determined by the relationship of the core levels involved, regardless of the energy of the excitation light and cannot be varied. As with photoelectrons, the wave function of Auger electron is also important for atomic resolution hologram analysis. Broadly speaking, Auger electrons mainly have components of the wave function of the orbital quantum number $l+1$ when the initial state excited has an orbital quantum number l. In the case of $KL_{2,3}L_{2,3}$ Auger, the core hole of 1s is filled with $2p$ electrons, and the Auger electron with orbital quantum number $l = 2$ is mainly emitted. The angular momentum of Auger electron also affects the hologram.[8]

Although the kinetic energy of Auger electrons is constant, there is a measurement method making good use of this feature. For example, diffraction spectroscopy combining Auger electron hologram and X-ray absorption spectroscopy is useful for investigating the magnetic structure of each atomic layer on the surface.[9]

2.1.3.4. *Inelastic mean free path*

The electron absorption cross section and inelastic mean free path (IMFP) show roughly the similar energy dependence regardless of the

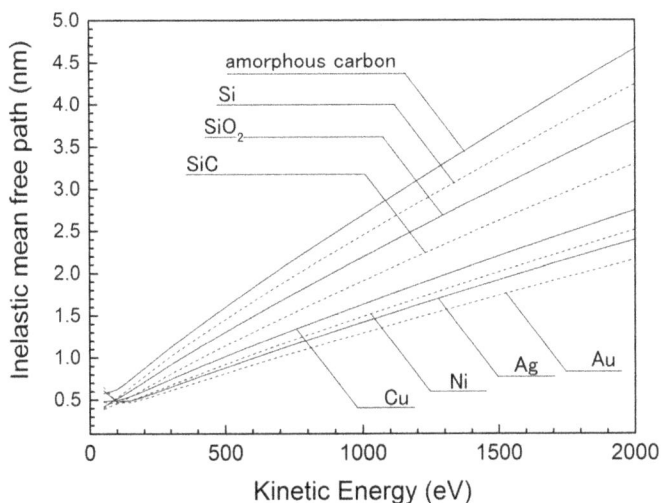

Fig. 2.1.4   Energy dependence of inelastic mean free path (IMFP) of various materials.

composition of the material. This is because photoelectrons in the solid mainly interact with valence electrons and are inelastically scattered by the excitation from the valence band to the conduction band or plasmon excitation. Photoelectrons of kinetic energy of 50 eV have the largest absorption cross section. Electron from deep several nm (tens of atomic layers) rarely reach the surface. The IMFP of photoelectrons with larger energy above 1 keV or smaller energy below 5 eV becomes longer. More precisely, IMFP can be estimated by TPP2M formula released by Tanuma et al.[10–12] The IMFP of a typical substance is shown in Fig. 2.1.4.

2.1.4.   Photoelectron Scattering

Electrons are scattered by the Coulomb potential and exchange inter-action of atoms. Some electrons are inelastically scattered and lose energy in solids, but the other are likely to be elastic scattered. This elastic scattering forms a photoelectron hologram. Figure 2.1.5 shows a simulation of the photoelectron scattering process. When electrons are scattered by atoms, the scatterer atom behave like convex lenses. The electron wave that has passed through the atom converges and generates strong intensity peak structure in the angular distribution at the direction of scatterer atom, called forward focusing peak. Here

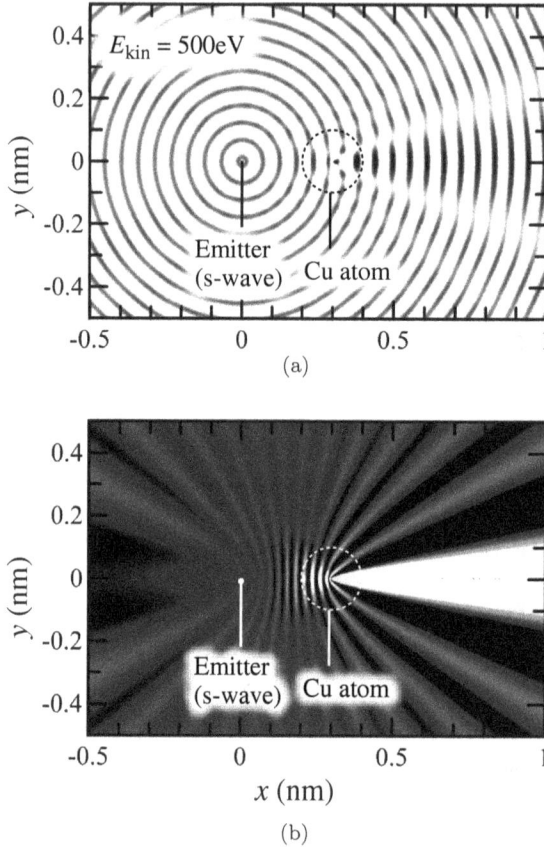

Fig. 2.1.5   Simulation of the photoelectron scattering process. The photoelectron with kinetic energy $E = 500$ eV and spherical s-wave is illustrated. The wave is scattered by the Cu atom located at 0.3 nm. (a) The real part of the wave function (b) The density of the photoelectron.

we introduce a partial wave expansion method, which is a method to numerically calculate only the wave function near the atom and connect it to analytic form of free electron wave functions. First, let's consider the wave function of free electrons with positive kinetic energy E. Then we solve the Schrödinger equation with $V(r) = 0$ in polar coordinate system.

$$\left[-\frac{\hbar^2}{2m}\Delta + V(r) \right] \psi(k, \mathbf{r}) = E\psi(\mathbf{r}). \qquad (2.1.4)$$

This solution is given by a linear combination of basis functions of outward expanding wave and inward converging wave.

$$\psi_{lm}^{+}(k, r) = (i + 1)^l k h_l^{(1)}(kr) Y_{lm}(\theta_r, \phi_r), \qquad (2.1.5)$$

$$\psi_{lm}^{-}(k, r) = (i + 1)^l k h_l^{(2)}(kr) Y_{lm}(\theta_r, \phi_r), \qquad (2.1.6)$$

where l, m is the azimuthal quantum number, magnetic quantum number, and $r = |\mathbf{r}|$. $h_l^{(1)}(x)$ and $h_l^{(2)}(x)$ are the spherical Hankel functions of the first kind and the second kind, respectively. For example, when $l = 0$, $m = 0$, this formula gives a simple spherical wave (s-wave), $\psi_{00}^{\pm}(r) = \exp(\pm ikr)/r$. When $l = 1$, $m = 0$, it becomes a wave like the p_z atomic orbital.

Let's consider the case where the s wave converges inward towards the origin as shown in Fig. 2.1.6(a). It passes through the origin and becomes a s wave spreading outward. The amplitude does not change. Similarly for the p_z wave, when the inward wave localized in the z axis passes through the origin, it becomes the outward expanding wave of the same angular distribution (Fig. 2.1.6(b)). The angular distribution $Y_{lm}(\theta, \phi)$ of the wave does not change even after passing through the origin.

Next, let's examine the spherical potential $V(r)$ created by the atoms at the origin. a is the interaction radius of $V(r)$. Its external potential is set to 0 (Fig. 2.1.6(c)). The inner converging wave coming from the outside spread out after entering the atom. Since the electron energy does not change before and after scattering, the wave number does not change. Because of the isotropic potential of the sphere, the angular distribution of the wave does not change. Since the number of electrons also does not change, the amplitude of the wave function also does not change. Only the phase of the wave changes. Phase shift is very important in scattering calculations. The phase is obtained by smoothly connecting the wave function between the wave within the interaction radius ($r < a$) and the external free electron wave function. The wave function $\psi_{lm}^{in}(r)$ in the interaction region is given by numerically solving the Schröedinger equation (2.1.4). The external wave function of free electrons is given by superposition of inward wave and outward wave as;

$$\psi_{lm}^{out}(k, r) = \psi_{lm}^{-}(k, r) + e^{2i\delta_l} \psi_{lm}^{+}(k, r). \qquad (2.1.7)$$

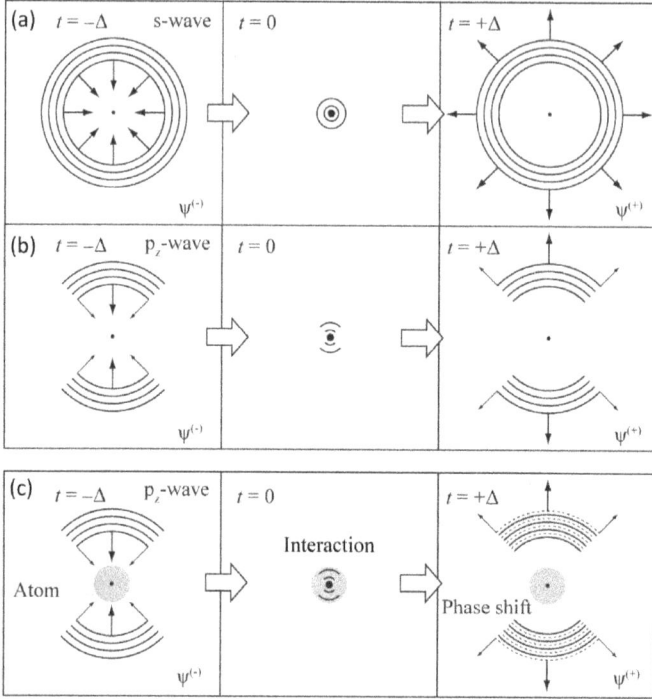

Fig. 2.1.6 Photoelectron wave scattered by atoms. Wave motion is calculated for each different angular momentum component. The converging wave turns to the expanding wave.

The phase shift δ_l is determined so that the wave functions inside and outside the interaction region are smoothly connected.

Here, we place the emitter atom at the origin. The wave function of photoelectrons is given by

$$\varphi_L(k, \mathbf{r}) = \sum_{lm} i^{l+1} k n_{Llm} h_l^{(1)}(kr) Y_{lm}(\theta_r, \phi_r), \qquad (2.1.8)$$

where L is the label for the emitted wave corresponding to the different orbital magnetic quantum numbers of the degenerated core-level state. The coefficient n_{Llm} is determined by the dipole transition matrix.

We shift the origin of the coordinate system to the scatterer atom located at $\mathbf{a}$. The wave function of the photoelectron is given by $\varphi_L(k, |\mathbf{r} + \mathbf{a}|)$. Using the partial wave expansion, this function in $r < a$ is expressed as

$$\varphi_L(k, |\mathbf{r} + \mathbf{a}|) = k \sum_{l,m} \frac{1}{2} C_{Llm}(\mathbf{a}) i^{l+1} \left[h_l^{(1)}(kr) + h_l^{(2)}(kr) \right] Y_{lm}(\theta, \phi),$$

(2.1.9)

$$C_{Llm}(\mathbf{a}) = \frac{(-i)^{l+1} \int \int \varphi_L(k, |R + a|) Y_{lm}^* \sin\theta d\theta d\phi}{j_l(kR)}.$$

In this equation, the coefficients of the inward converging wave $h^{(2)}$ and the outward expanding wave $h^{(1)}$ is same. In the scattering process, the inward wave enters the atom and exits as an outward wave, its phase changes. Therefore, the wave function of the scattering state is given by

$$\hat{\varphi}_L(k, |\mathbf{r} + \mathbf{a}|) = k \sum_{l,m} \frac{1}{2} C_{Llm}(\mathbf{a}) i^{l+1} [e^{2i\delta_l} h^{(1)}(kr) + h^{(2)}(kr)] Y_{lm}(\theta, \phi).$$

(2.1.10)

The scattered wave from this equation is derived as

$$\psi_L(k, \mathbf{r}, \mathbf{a}) = \sum_{l,m} \frac{1}{2} C_{Llm}(\mathbf{a}) i^{l+1} (e^{2i\delta_l} - 1) h^{(1)}(kr) Y_{lm}(\theta, \phi). \quad (2.1.11)$$

The waves shown in Fig. 2.1.5 was calculated using this method.

Photoelectrons are scattered by many atoms. Since photoelectrons are detected at a distance far from the excited atoms, the intensity of photoelectrons, *i.e.* photoelectron holograms, is given by

$$I(\mathbf{k}) = \sum_L \left| \varphi_L(\mathbf{k}) + \sum_i \psi_L(\mathbf{k}, \mathbf{a}_i) \right|^2. \quad (2.1.12)$$

Here, we redefined the wave function as

$$\varphi_L(\mathbf{k}) \equiv \lim_{r \to \infty} \varphi_L(k, \mathbf{r}), \quad (2.1.13)$$

$$\psi_L(\mathbf{k}, \mathbf{a}_i) \equiv \lim_{r \to \infty} \psi_L(k, \mathbf{r}, \mathbf{a}_i). \quad (2.1.14)$$

The Cu *LMM* Auger electron hologram simulated by the above equation is shown in Fig. 2.1.7. Forward focusing peaks appear in the direction of {110} and {100} in face-centered-cubic lattices. Simulation of the Auger electron hologram explains the experiment well.

Cu Auger $E_{kin} = 914eV$

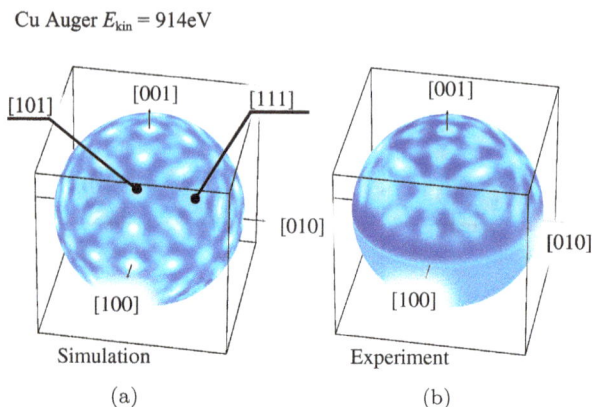

Simulation   Experiment

(a)   (b)

Fig. 2.1.7   (a) Simulation of an LMM Auger electron hologram (kinetic energy 914 eV) from copper (Cu) single crystal. An atomic cluster with a radius of 0.9 nm was used. (b) Experimental *LMM* Auger electron hologram obtained. In the experiment, only the hologram above the surface can be measured, so the data is limited to the upper half of the sphere.

In actual experiments, holograms can be observed under various conditions. For example, a negative pattern similar to a photoelectron hologram whose contrast is inverted also appears in the energy loss electrons.[13]

2.1.5.   Atomic Image Reconstruction

Three-dimensional atomic arrangement information is involved in the hologram, and it is necessary to reconstruct a three-dimensional atomic image from this information. In the case of a photoelectron hologram, an atomic image can not be obtained by a calculation method based on Fourier transform such as fluorescent X-ray holography. In the forward scattering direction, the phase of the electron wave changes greatly, and the scattered wave is not a simple *s*-wave. Therefore, an algorithm with "scattering pattern function" as the basis function has been developed.[1,2]

2.1.5.1. *Analysis using scattering pattern function*

The hologram equation suggests that the hologram is regarded as the sum of the scattering patterns formed by each surrounding atom. The scattering pattern function is defined as,

$$t(\mathbf{k}, \mathbf{a}) = |\mathbf{a}| \sum_L \left[2Re[\varphi_L^*(\mathbf{k})\psi_L(\mathbf{k}, \mathbf{a})] + |\psi_L(\mathbf{k}, \mathbf{a})|^2 \right]. \qquad (2.1.15)$$

In this equation, the coefficient $|\mathbf{a}|$ is introduced to correct scattering intensity weakening which increases with the atomic distance between the emitter atom and the scatterer atom. Next, a three-dimensional function $g(\mathbf{a})$ representing the distribution of atoms is defined. If the three-dimensional function is expressed as

$$g(\mathbf{a}) = \sum_h \frac{\delta(\mathbf{r} - \mathbf{a}_h)}{|\mathbf{a}_h|}, \qquad (2.1.16)$$

the hologram is given by the following integral,

$$\chi(\mathbf{k}) = \int g(\mathbf{a})t(\mathbf{k}, \mathbf{a})\mathrm{d}\mathbf{a}. \qquad (2.1.17)$$

The scattering pattern function $t(\mathbf{k}, \mathbf{a})$ is six dimensions, *i.e.* position of scattered atoms, kinetic energy, and detected orientation. It also reflects the shape of the direct wave $\varphi_L(\mathbf{k})$ determined by the transition matrix elements of photoelectrons and Auger electrons. Figure 2.1.8 shows an example of the scattering pattern function when *s*-waves with kinetic energy of 914 eV are scattered by Cu atoms. Figure 2.1.8(a) shows scattering atoms located at $z = 0.3$ nm. A strong peak on the z-axis is the forward focusing peak, and ring-like diffraction patterns appear around it. Figure 2.1.8(b) is for $z = 0.9$ nm. As the position of the atom gets farther, the diffraction ring spacing decreases, *i.e.* the spatial frequency in wave space increases. The forward focused peak direction represents the direction of the scatterer atom and the spatial frequency represents the interatomic distance. In this way, three-dimensional information of the atomic arrangement is recorded on the hologram.

In an actual hologram, atoms at various positions form corresponding scattering patterns, and these ensembles are observed as photoelectron holograms. If the photoelectron hologram can be decomposed into scattering patterns generated from individual atoms, the atomic position can be obtained.

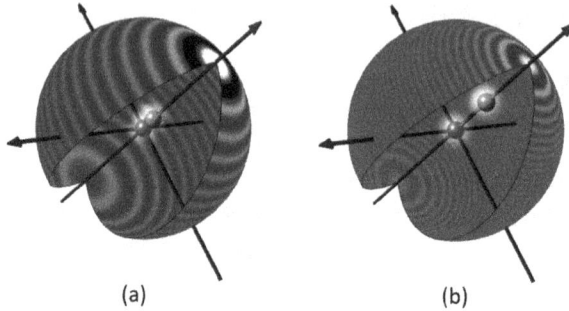

Fig. 2.1.8   Examples of scattering pattern functions. The spherical wave (s-wave) with $E = 914\,\mathrm{eV}$ is scattered by copper atoms. (a) The atomic distance between photoelectron emitting atom and scatterer atom is 0.3 nm. (b) The case of the atomic distance for 0.9 nm.

2.1.5.2. *Reconstruction of atomic images using scattering patterns*

In order to obtain the atomic distribution function $g(\mathbf{a})$ from the photoelectron hologram $\chi(\mathbf{k})$ using the relation of eq. 2.1.17, $g(\mathbf{a})$ is described by introducing a voxel which divides the real space into a three-dimensional lattice. These voxels can be made independently of the crystal lattice structure. Here, the j-th voxel value at the position of $\mathbf{a}_j$ is defined as g_j. Also, the value of the hologram in the wave vector of $\mathbf{k}_i$ is defined as χ_i. Here, i is an index of the measurement point. Then the expression (2.1.17) is given by

$$\chi_i = \sum_j^N g_j t(\mathbf{k}_i, \mathbf{a}_j) \Delta V, \qquad (2.1.18)$$

where ΔV is the volume of the voxel. This equation is a linear simultaneous equation for converting from a three-dimensional real space voxel to a three-dimensional hologram. However, this simultaneous equation can not be solved by a simple gradient method because the total number of "unknowns" is overwhelmingly larger than the number of data of holograms. For example, a hemisphere of a single energy hologram having a solid angle resolution of 1° contains about 20,000 measurement points (M). When dividing a real space of ±1 nm with a resolution of

0.01 nm, the number of voxel elements (N) is 200^3. The value of the voxel in the real space corresponds to the "unknown" of the equation.

Next, we define the information entropy S by the following equation and solved using the maximum entropy method.

$$S = -\sum_{j}^{N} g_j^{(n)} \ln \frac{g_j^{(n)}}{g_j^{(n-1)}} - \lambda C, \qquad (2.1.19)$$

$$C = \frac{1}{M} \sum_{i}^{M} \frac{\| \chi_i^{\mathrm{exp}} - \chi_i^{\mathrm{cal}} \|^2}{\sigma_i^2} - 1 \qquad (2.1.20)$$

n is the number of iterations. χ^{exp} and χ^{cal} is the experimental and calculated hologram, respectively. σ is the standard deviation of the noise. Atomic images are obtained by maximizing information entropy. This calculation method is called SPEA-MEM (Scattering Pattern Extraction Algorithm with Maximum Entropy Method). Fig. 2.1.9 is an image reconstructed by SPEA-MEM from Cu Auger electron hologram (Fig. 2.1.7).

2.1.5.3. *Visualization of graphite atom arrangement*

Figure 2.1.10(a) shows a photoelectron hologram from graphite. The characteristic six-fold symmetric arc patterns are in-plane CC coupled diffraction rings. Figure 2.1.10(b)–(f) shows real space atomic arrangement of graphite reconstructed with SPEA-MEM.[14] In the case of graphite with an AB layered structure, since there are four atomic sites in the unit cell which are not equivalent, the actually measured pattern is the sum of hologram patterns of all sites. The reconstructed atomic image is a superposition of four kinds of lattices. The layered structure of graphite appears in the cross section in the interlayer direction. Of course, such a layered structure does not appear in monolayer graphene.[15, 16] By using SPEA-MEM and an appropriate scattering pattern matrix, 3D atomic arrangement can be reconstructed.

Furthermore, since the phase shift depends on the atomic number, the scattering pattern differs from element to element. If the atomic number differs by a factor of two or more, not only the atomic

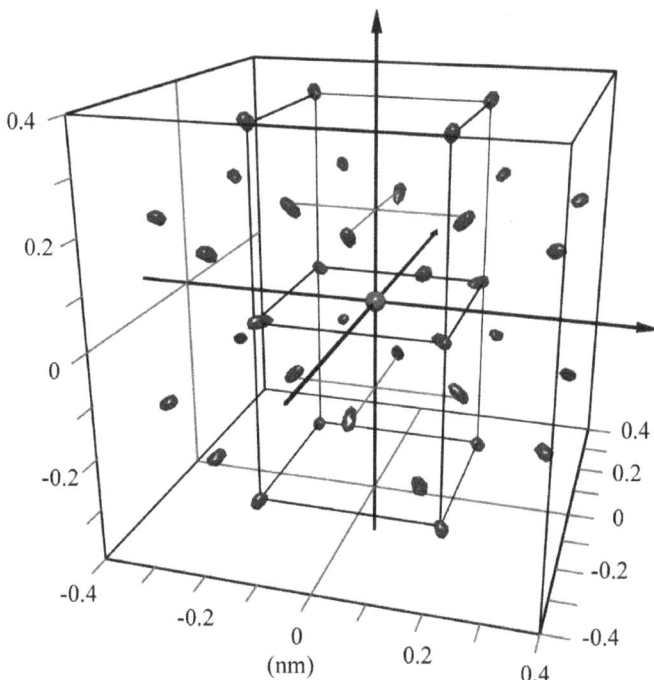

Fig. 2.1.9 Three-dimensional atomic image reconstructed from Cu Auger electron hologram.

arrangement but also the elements of the scatterer can be reconstructed. There is a successful example to reproduce the InP crystal lattice.[17]

2.1.5.4. *Circularly polarized light excitation and atomic stereogram*

Circularly polarized light is an electric field corresponding to $p_{\pm1}$ spherical harmonic functions $|1, \pm1\rangle$. For light propagating along the z axis, the phase difference between the x polarization component and the y polarization component is $\pm90°$. On the xy plane of the right-handed system, circularly polarized light having an electric field vector e rotating along the z-axis from the x-axis to the y-axis direction (clockwise viewed from the light source side) is $\sigma = +1$. $\sigma = -1$ circularly polarized light rotates counterclockwise. These are also referred to as left circular

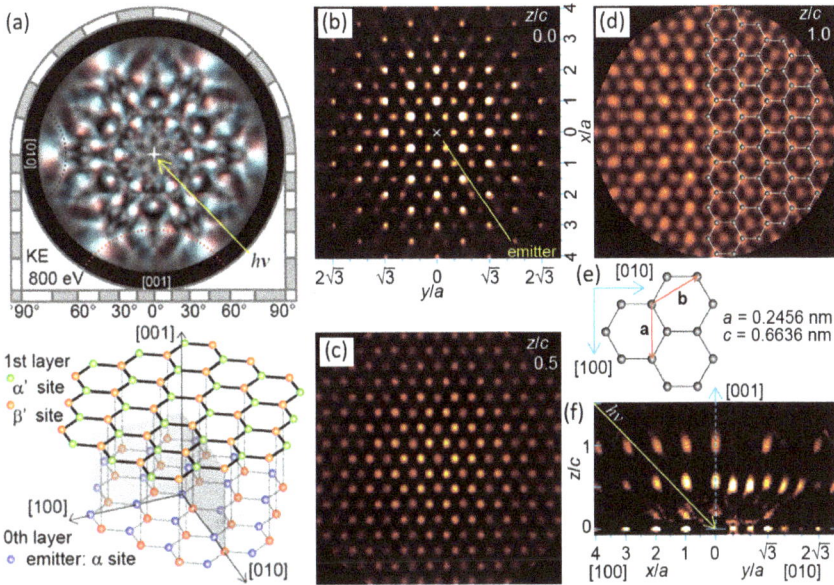

Fig. 2.1.10 (a) Photoelectron angular distribution of C 1s core level of graphite. Real space image reconstructed by photoelectron holography of, (b) horizontal cross section including the emitter atom site, (c) one layer above, and (d) two layer above the emitter atom site, (f) vertical cross section. Images from two types of sites are overlapped.

polarization and right circular polarization. Be careful when inverting it when viewed from the downstream.

The angular momentum of the entire system is preserved before and after the photo-excitation process. In the excitation process, the photoelectron angular momentum increases or decreases ($\Delta m_l = \pm 1$) corresponding to the excitation light helicity $\sigma = \pm 1$. Figure 2.1.11 shows the wave function when circularly polarized light is irradiated. Circularly polarized light enters perpendicularly to the paper surface. When the $p_{m=+1}$ state is excited, the final state becomes $d_{m=+2}$. Since this wave function has a spiral wave front, the forward focusing peak and interference fringes are shifted in the direction of helicity when scattered by surrounding atoms. This phenomenon was used in the measurement of atomic stereo photographs.[18]

Figure 2.1.12 is an example of photoelectron diffraction from InSb (001). In the case of In $3d$, the first nearest Sb atom exists in the [111] direction, and a bright forward focusing peak is observed in the

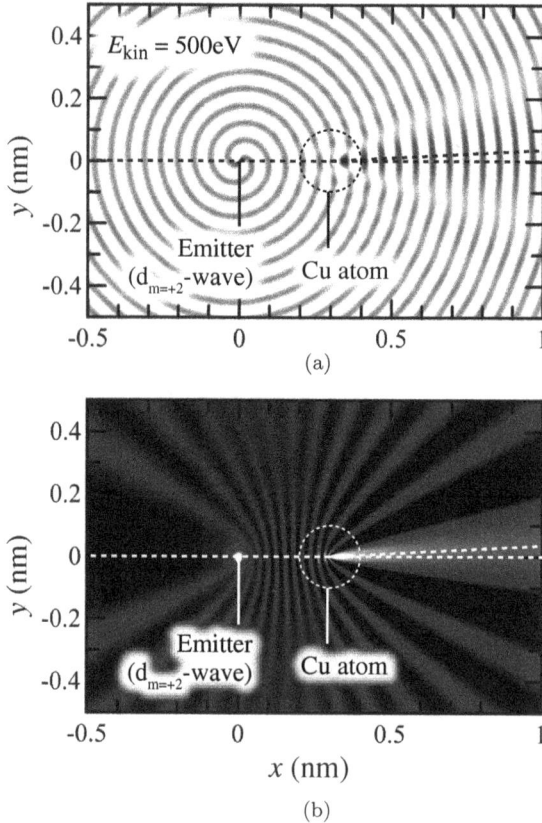

Fig. 2.1.11 (a) Wave function (real part) when p electron is excited by circularly polarized light and scattered by a Cu atom. (b) Probability density of electron.

corresponding direction. In the case of Sb 3d, the forward focusing peak of the first nearest In atom appears in the direction of [$\bar{1}$11]. When circularly polarized light is used, the forward focusing peaks show parallax angle shift. The angular shift $\Delta\phi$ of the forward focusing peak due to atoms in the interatomic distance R is expressed by the following equation.

$$\Delta\phi = \frac{m^*}{kR\sin^2\theta} \qquad (2.1.21)$$

k is the wave number of the photoelectron, and θ is the angle formed between the incident light and the direction of the photoelectron

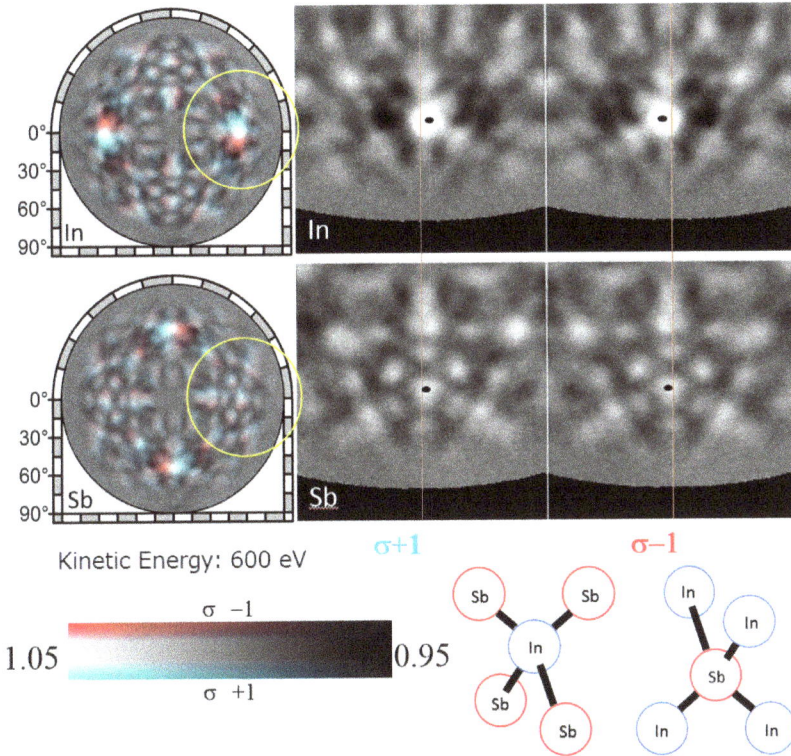

Fig. 2.1.12 Photoelectron holograms of In $3d$ and Sb $3d$ from InSb (001) surface. Both kinetic energy is 600 eV. The yellow circles indicate the [111] direction.

emission. The angular momentum of photoelectron m^* is the sum of the angular momentum m_i of the excited orbital and the polarized light σ of light, depending on θ. When comparing the forward focusing peaks in the [111] direction, since In atom is seen from Sb atom three times farther than Sb atom seen from In atom, the parallax angle is also as small as that from In atom. Because the forward focusing peaks from different sites appear in different directions, the geometric sites of photoelectron emitting atoms can be specified. For example, core level photoelectron diffraction of the dopant atoms can be used to determine whether the dopant occupies the A site or the B site compared to that of the A and B atoms. The element-selective density distribution can be obtained by adapting the above approach to the valence band.

2.1.6. Summary: Development of Photoelectron Holography

Finally, we summarize the point to keep in mind for photoelectron hologram studies. Since the amplitude of the photoelectron hologram is several hundredths to several tenths of the intensity of the reference wave, the hologram measurement of the bulk crystal is relatively easy. On the other hand, in order to obtain a hologram of impurities, it is important to measure with sufficient signal-to-noise ratio, paying attention to the influence of signal of the substrate crystal. It is desirable that kinetic energy of photoelectrons is 400 eV or more. Photoelectrons with large kinetic energy have short wavelengths and have good spatial resolution. In addition, it is necessary to accurately consider the refractive effect of the surface due to the inner potential. However, for surface structure studies, it is important to remember that as the kinetic energy increases the sensitivity to the surface decreases. To use the holographic reconstruction algorithm SPEA-MEM, even a single energy is sufficient, but it is necessary to measure the two-dimensional angular distribution of the electron intensity over the solid angle of 1π sr (steradian). At least $1°$ is required for angular accuracy. Furthermore, it is necessary to pay attention to the orbital quantum number of the excited core level. When the number of orbital quantum numbers in the final state is large, there are many nodes in the reference wave, so the interference fringes become weak. Therefore, hologram reconstruction from the s orbital core level tends to increase accuracy, while the accuracy of the f orbital case tends to decline.

Recently, technologies such as microscopic photoelectron holography using a convergent photon beam, photoelectron diffraction microscope, holographic spectroscopy combined with various spectroscopic methods, and ultrafast photoelectron holography by a pump probe method are progressing rapidly. It is also being applied to new functional materials.

References

[1] T. Matsushita *et al.*, *Phys. Rev. B* **75**, 085419 (2007).
[2] T. Matsushita *et al.*, *Phys. Rev. B* **78**, 144111 (2008).

[3] D. E. Eastman *et al.*, *Nucl. Instrum Meth.* **172**, 327 (1980).

[4] H. Daimon, *Rev. Sci. Instrum.* **59**, 545 (1988).

[5] H. Matsuda *et al.*, *Phys. Rev. E* **71**, 066503 (2005).

[6] B. Krömker *et al.*, *Rev. Sci. Instrum.* **79**, 053702 (2008).

[7] G. Öhrwall *et al.*, *J. Electron Spectrosc. Relat. Phenom.* **183**, 125 (2011).

[8] F. Matsui *et al.*, *Phys. Rev. Lett.* **114**, 015501 (2015).

[9] F. Matsui *et al.*, *Phys. Rev. Lett.* **100**, 207201 (2008).

[10] S. Tanuma, C. J. Powell, and D. R. Penn, *Surf. Interface Anal.* **21**, 165 (1994).

[11] S. Tanuma, C. J. Powell, and D. R. Penn, *Surf. Interface Anal.* **43**, 689 (2011).

[12] A. Jabłoński and C. J. Powell, *J. Electron Spectrosc. Relat. Phenom.* **100**, 137 (1999).

[13] F. Matsui *et al.*, *J. Phys. Soc. Jpn.* **81**, 013601 (2012).

[14] F. Matsui *et al.*, *J. Phys. Soc. Jpn.* **81**, 114604 (2012).

[15] S. Roth *et al.*, *Nano Lett.* **13**, 2668 (2013).

[16] H. Matsui *et al.*, *Surf. Sci.* **635**, 1 (2015).

[17] T. Matsushita *et al.*, *J. Phys. Soc. Jpn.* **82**, 114005 (2013).

[18] H. Daimon, *Phys. Rev. Lett.* **86**, 2034 (2001).

Subchapter 2.2

X-RAY FLUORESCENCE HOLOGRAPHY

Kouichi Hayashi

Nagoya Institute of Technology

X-ray fluorescence holography is a technique for recording three-dimensional arrangement of atoms using the interference of fluorescent X-rays or incident X-rays around emitter atoms in single crystal samples. In particular, it is effective to analyze local structures around dopants. Since atomic images can be reconstructed over a range of 2 to 3 nm, the X-ray fluorescence holography makes us possible to analyze lattice distortion and special clusters around dopants. In this section, we will describe the basic principle of the method and introduce two applications.

2.2.1. Introduction

Fluorescent X-ray is radiation from an atom when its inner shell is excited by a primary X-ray then it transits to the ground state. X-ray fluorescence itself has been used for elemental analysis for a long time since its wavelength is specific value depending on the element. There are wide applications such as material composition analysis, trace element analysis in biological samples, product management, and clarification of origins in archeology owing to non-destructive testing without troublesome vacuum system.

On the other hand, a structural analysis technique utilizing the diffraction phenomenon of X-ray fluorescence has also existed. For example, the Kossel line,[1] which is a two-dimensional pattern formed by Bragg reflection of fluorescent X-rays in a single crystal, had been used for evaluation of crystallinity in the past. Since the Kossel line was discovered in 1935, it can be said to be a classic technique of a century ago. After a suggestion by Szökez, we understand that the Kossel lines are interpreted as parts of an X-ray fluorescence hologram, which records atomic arrangement of atoms. After 1986, the research of X-ray fluorescence holography rapidly progressed with the development of the X-ray detectors.

Photoelectron holography, which uses the same principle of X-ray fluorescence holography, can measure easily the holograms because the amplitude of the photoelectron intensity variation is 20–50% of the background intensity as described in previous section. Therefore, its demonstration was performed in 1990[2] earlier than X-ray fluorescence holography. On the other hand, X-ray fluorescence holography with the amplitude of only about 0.1% was demonstrated by Tegze and Faigel in 1996.[3] In their experiment using a conventional X-ray tube, it took three months of measurement time to earn sufficient statistical accuracy. For this reason, high flux monochromatic X-rays are necessary, and therefore the activity of X-ray fluorescence holography immediately shifted to synchrotron radiation.

By combining brilliant X-ray source and advanced X-ray detection system, measurement time in three months was reduced to several hours. In addition, the resolution of the reconstructed atomic image reaches 0.5 Å,[4] and light elements such as oxygen can be visualized.[5] Since the samples are required orientational symmetry, amorphous or powder samples cannot be measured, and single crystals and epitaxial films with the sizes of several millimeters are targeted. However, not only a crystal with a long range translational order but also clusters and surface adsorbates, dopants,[6] quasicrystals[7] are also applicable.

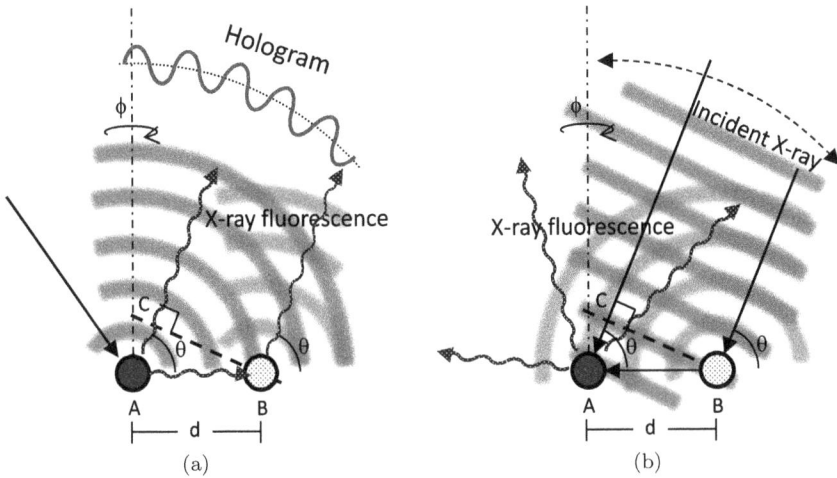

Fig. 2.2.1   Principle of X-ray fluorescence holography using dimer model. (a) normal mode. (b) Inverse mode. This figure is taken from Ref. [8].

2.2.2.   Principle

2.2.2.1.   *Normal and inverse modes*

There are two modes of X-ray fluorescence holography: normal and inverse modes. Figure 2.2.1 illustrates both the modes using a simple dimer model. In the normal mode in Fig. 2.2.1(a), atom A is excited by incident X-rays to emit fluorescent X-rays. A part of the wave of the fluorescent X-ray is scattered by an adjacent atom, then interferes with non-scattered wave. These scattered waves and non-scattered waves play the role of an object wave and a reference waves, respectively, in holography. Therefore, the hologram is just a two-dimensional spatial distribution of the X-ray fluorescence intensities.

In the case of the dimer, the hologram pattern can be represented by a simple formula. The path difference between the object and the reference waves can be expressed as $d(1 - \cos\theta)$, where d is the distance between the atoms A and B and θ is the angle between AC and AB. When the fluorescent X-ray is scattered by the atom, the phase of the wave shifts by π. When $d(1 - \cos\theta)/\lambda$ is a half integer, intensity maxima appear, where λ is the wavelength of the fluorescent X-ray. Using the scattering factor of atom A, $f\,(\theta, \lambda)$, the X-ray fluorescence intensity $I(\theta, \lambda)$ can be expressed as

$$I(\theta, \lambda) = \left| 1 - \frac{\lambda r_e f(\theta, \lambda)}{2\pi d} e^{i2\pi d(\cos\theta - 1)/\lambda} \right|^2$$

$$= 1 - 2\mathrm{Re}\left(\frac{\lambda r_e f(\theta, \lambda)}{2\pi d} e^{i2\pi d(\cos\theta - 1)/\lambda} \right)$$

$$+ \left| \frac{\lambda r_e f(\theta, \lambda)}{2\pi d} e^{i2\pi d(\cos\theta - 1)/\lambda} \right|^2, \qquad (2.2.1)$$

where r_e is the classic electron radius. Since the cross section of the electron for X-ray is very small, $\lambda f(\theta, \lambda)/2\pi d$ is less than 10^{-3}. Thus, Eq. (2.2.1) can be approximated to

$$I(\theta, \lambda) \cong 1 - 2\mathrm{Re}\left(\frac{\lambda r_e f(\theta, \lambda)}{2\pi d} e^{i2\pi d(\cos\theta - 1)/\lambda} \right). \qquad (2.2.2)$$

Here, the second term means the hologram recording the position of atom B.

Figure 2.2.1(b) shows the principle of the inverse mode. Since the inverse mode is a time-reversed version of normal mode, basically it is possible to record a hologram equivalent to normal mode.[9] Here, the wave of the incident X-ray directly approaching the atom A is the reference wave, and the wave approaching the atom A after being scattered by the atom B plays a role as the object wave. The reference and object wave form X-ray standing wave around the atom A. The pattern of the X-ray standing wave varies depending on the direction of the incident X-ray, and consequently changes the intensity of the X-ray fluorescence from the atom A. The variation of the X-ray fluorescence intensity as a function of the orientation of the incident X-ray beam becomes a hologram equivalent to the normal mode. Therefore, Eqs. (2.2.1) or (2.2.2) is also applied to the inverse mode. However, in that case, θ is the polar angle of the incident X-ray, and λ is the wave length of the incident X-rays.

As described above, since the inverse X-ray fluorescence holography uses the incident X-ray as a wave of holography, it can record the hologram with all the wavelengths (energies) of the X-rays less (more) than the absorption edge of the atom A. This point is quite different from the normal mode in which the hologram can be recorded only at the wavelength of the fluorescent X-rays. By recording a hologram with many wavelengths, more accurate atomic images can be obtained due to the suppression of the ghost images. For this reason, most of X-ray fluorescence holography experiments have been carried out in inverse mode.

For a more realistic calculation model with a number of atoms, it is more convenient to use Eq. (2.2.3), which is an extension of Eq. (2.2.2). The intensity $I(\mathbf{k})$ of the X-ray fluorescence can be expressed as follows,

$$I(\mathbf{k}) \cong 1 - 2\mathrm{Re} \sum_{j} \left[\frac{r_e f_j(\theta_{\mathbf{r}_j}^{\mathbf{k}})}{r_j} e^{i(-\mathbf{k}\cdot\mathbf{r}_j - kr_j)} \right] + \left| \sum_{j} \frac{r_e f_j(\theta_{\mathbf{r}_j}^{\mathbf{k}})}{r_j} e^{i(-\mathbf{k}\cdot\mathbf{r}_j - kr_j)} \right|^2,$$

(2.2.3)

where $\mathbf{r}_j$ is the coordinate of the jth atom, and f_j is the atomic scattering factor of the jth atom. $\theta_{\mathbf{r}_j}^{\mathbf{k}}$ is the angle between $\mathbf{k}$ and $\mathbf{r}_j$. Eq. (2.2.3) is used for the inverse mode, but it can be used in normal mode by replacing $\mathbf{k}$ with $-\mathbf{k}$. The second term expresses holographic oscillation like Eq. (2.2.1). The third term is negligible because of a very small value except when it satisfies the Bragg condition.

2.2.2.2. *Atomic image reconstruction and multiple-wavelength recording*

In the case of optical holography, an image of the original object is formed by irradiating a reference wave on the film recording the hologram. However, X-ray fluorescence holography cannot reconstruct an atomic image by such a way, but the atomic image can be obtained by applying Fourier transformation to the hologram using a computer.

The following Eq. (2.2.4) is Helmholtz-Kirchhoff formula,[10] namely, a kind of three-dimensional Fourier transformation.

$$U(\mathbf{r}) = \iint e^{-ikr} \chi(\mathbf{k}) d\sigma. \qquad (2.2.4)$$

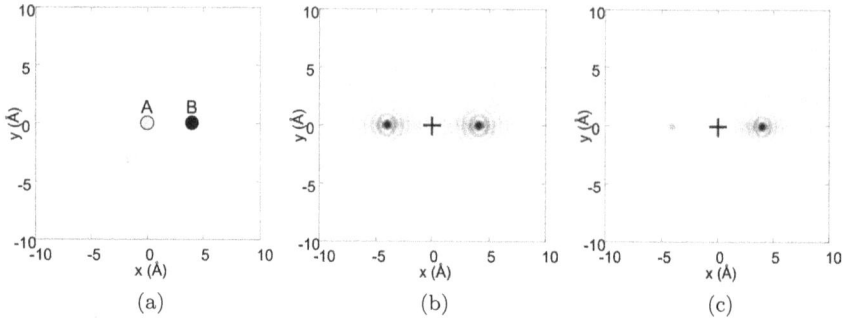

Fig. 2.2.2   Multiple-wavelength reconstruction (a) A dimer model. (b) Image from single-wavelength hologram. (c) Image from multiple-wavelength hologram. This figure is taken from Ref. [8].

Here, χ (k) is a hologram, which is the oscillation component extracted from I(k). Although this formula can be used to reconstruct atomic images from holograms, we use the following equation[11] when reconstructing from multiple-wavelength holograms.

$$U_k(\mathbf{r}) = e^{ikr} \iint e^{-ikr} \chi(\mathbf{k}) d\sigma. \qquad (2.2.5)$$

Although this formula is just obtained by multiplying Eq. (2.2.4) by e^{ikr}, only the true atomic images are emphasized and the ghost images are diminished by summing up $U_k(\mathbf{r})$ at different wavelengths. The principle of the multiple-wavelength reconstruction can be explained using the reconstructions of the dimer in Fig. 2.2.2.

Figure 2.2.3(a) shows the X-ray fluorescence hologram calculated with the dimer model. Fluorescent X-rays are emitted from atom A, and atom B is a scatterer of the fluorescent X-rays. The wavelength of incident X-ray of 1.03 Å (energy: 12.0 keV) is used for the calculation of inverse mode. In the hologram in the wave vector space, many rings are formed around the k_x axis. By applying Fourier transformation to this pattern, the atomic image is obtained as shown in Fig. 2.2.2(b). The position of the atomic image is determined by the vector defined by connecting the centers of the rings (in this case, k_x) and the interval between the rings. The length of the interval is inversely proportional to the distance of the atomic image from the origin.

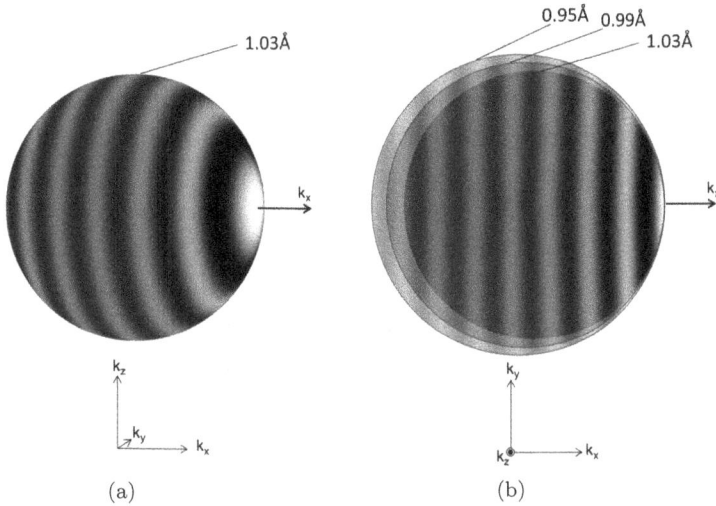

Fig. 2.2.3 Hologram formed by one scatterer. (a) Bird-eyes view of single wavelength hologram. (b) Top view of the overlapped three holograms recorded by different wavelengths. This figure is taken from Ref. [8].

Two atomic images are reconstructed at $x = \pm 4$ Å in Fig. 2.2.2(b), but the original position of the scatterer atom is only at $x = 4$ Å. It is called "twin image problem in holography" that a false atomic image appears at a centrosymmetric position with respect to the origin. The twin image problem causes deterioration of reconstructions, such as image disappearance, distortion and appearance of a ghost image. Therefore, the twin image problem must be solved, and multi-wavelength reconstruction method is most frequently used for its solution. Figure 2.2.3(b) shows a hologram pattern from the same dimer measured at three wavelengths of 1.03 Å (12.0 keV), 0.99 Å (12.5 keV) and 0.95 Å (13.0 keV). Here, the three holograms are overlapped such that their right ends are aligned. The top view shows that the stripe patterns of the three holograms coincide. In this manner, even changing the wavelength of the hologram, the interference pattern itself does not change and only the range of the hologram changes. Note that the phases of the patterns coincide only when aligning the edges of the holograms directing to the scatterer.

The multiple-wavelength reconstruction algorithm uses such a property of holograms. As shown in Fig. 2.2.3(b), if the right ends in the k_x direction are aligned and the entire holograms are Fourier-transformed all together, the image should be strongly reconstructed. However, when the left end is aligned, the phases do not coincide, and consequently the image intensity decreases. Figure 2.2.2(c) shows the atomic image reconstructed from the multiple-wavelength hologram. Here, in addition to the above three holograms, two holograms with wavelengths of 0.92 Å (13.5 keV) and 0.88 Å (14.0 keV) were added to emphasize the effect of the multiple-wavelength reconstruction (total 5 holograms). In the reconstruction, while the image at $x = 4$ Å is still strong, the one at $x = -4$ Å is greatly weakened. Although the effect of the multiple-wavelength method is remarkable for the calculated hologram of the dimer, the effect is also great for experimental data. In order to obtain quantitative physical information of local structure such as lattice distortion, at least ten multi-wavelength holograms are required.

On the other hand, if we measure multiple-wavelength holograms, much measurement time is necessary. For example, measurement time of single-wavelength hologram is typically three hours. In this case, the measurement of ten different wavelength holograms requires 30 hours. Therefore, a sufficient number of holograms might not be able to be measured at synchrotron radiation facilities where the beam time is limited. In order to solve such problems, Dr. Matsushita at SPring-8 developed an atomic image reconstruction algorithm called SPEA-MEM (scattering pattern matrix extraction algorithm using the maximum-entropy method).[12] SPEA-MEM can provide a real space image by fitting a calculated hologram without twin image.

2.2.3. Experimental Setup

Figures 2.2.4(a) and (b) show schematic drawings of the normal and inverse modes, respectively. In the normal mode, the spatial intensity distribution of the fluorescent X-rays is a hologram. Therefore, using a two-dimensional detector, the hologram can be measured without moving the sample and the detector. However, secondary radiations from the sample are often detected other than the wanted fluorescent X-rays. Since the two-dimensional detector cannot resolve different kinds of radiations, it is practically impossible to measure the hologram of

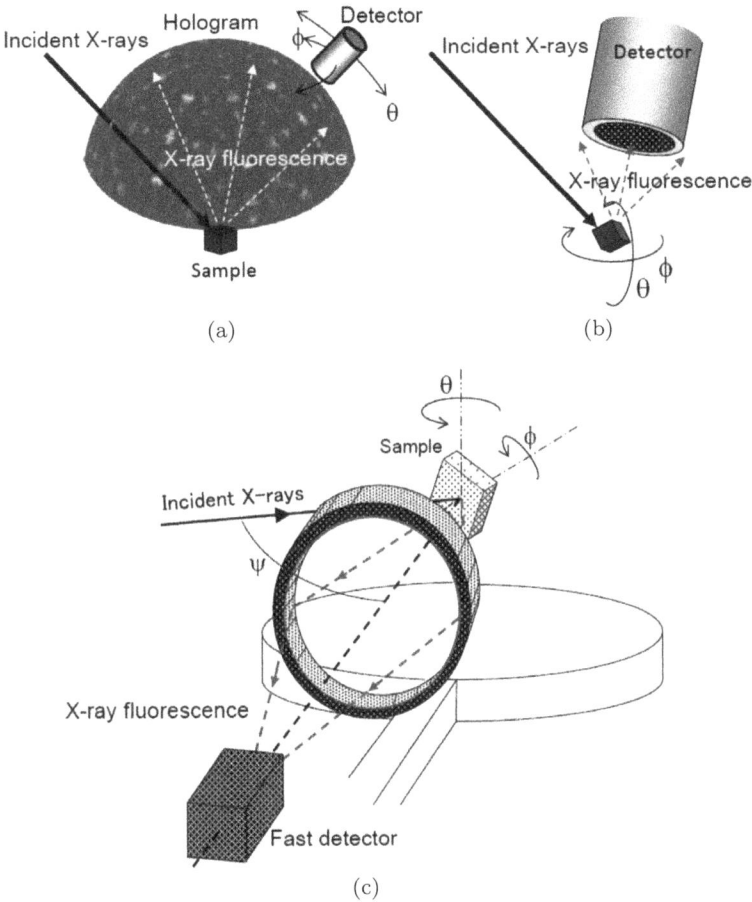

Fig. 2.2.4 Schematic drawing of the hologram measurement. (a) Normal mode. (b) Inverse mode. (c) Experimental setup consisting of cylindrical analyzer and fast X-ray detector. This figure is taken from Ref. [8].

only the target element. Therefore, we desire a detector with an energy resolution capable of selecting only one fluorescent line. Currently, there is no X-ray detector with an appropriate energy resolution. Therefore, hologram measurements have been basically performed by scanning a one-element semiconductor detector having high energy resolution around samples.[13]

As shown in Fig. 2.2.4(b), the hologram obtained in inverse mode can be measured by recording the X-ray fluorescence intensity while the sample orientation is changed two-dimensionally. Although all fluorescent X-rays from the sample should be detected ideally, it is practically difficult to carry out. Therefore, it is desirable to make the detector as close as possible and to detect with a solid angle as wide as possible. In many cases, we have used a detection system that combines a curved analyzing crystal and a fast X-ray detector of avalanche photodiode instead of a semiconductor detector, as shown in Fig. 2.2.4(c). In this case, although the detectable solid angle of the fluorescent X-rays is limited, the count rate of X-ray photons improves by more than three orders of magnitude. Therefore, this system makes us possible to obtain a high quality hologram within a few hours, and it has been frequently used at synchrotron radiation facilities where strong X-ray beam can be obtained.[14]

In order to avoid the sample shape effect affecting the intensity modulation of X-ray fluorescence, the sample surface should be polished to be flat. A millimeter sized sample is required so as to be irradiated by all flux of the incident beam even at higher θ. Using a micro focused X-ray beam, a micrometer-sized sample is possible to be applied in principle. In the measurement, the angle step of the azimuth angle ϕ and the X-ray incident angle θ should be less than 1°. In order to earn the k-space volume as much as possible, the scan angle ranges should be $0° \leq \phi \leq 360°$ and $0 \leq \theta \leq 80°$.

2.2.4.   Data Processing

2.2.4.1.   *Removal of component of normal mode*

The holographic oscillation with respect to the background can be extracted by the following equation:

$$\chi(\phi, \theta) = (I(\phi, \theta) - I_0(\vartheta))/I_0(\vartheta) \qquad (2.2.6)$$

Here, $I_0(\theta)$ is the average value of the X-ray fluorescence intensity at the angle θ. Figure 2.2.5(a) shows the resulted hologram pattern, and vertical stripes are seen. Using the above experimental arrangement with fixed exit angle of X-ray fluorescence, the resulting hologram pattern is the sum of one-dimensional normal hologram components and two-dimensional inverse hologram components. Stripe is due to normal

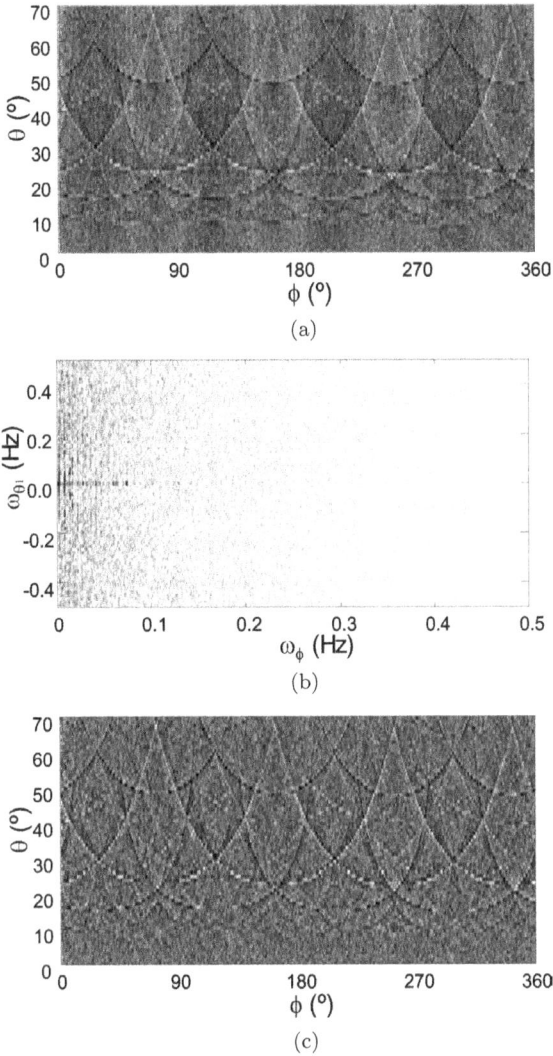

Fig. 2.2.5 Data processing for removing normal component. (a) Raw data. (b) Fourier-transform of (a). (c) Inverse Fourier-transform of (b) after removing the spots at $\omega_\theta = 0.0$.

component. This can be removed by Fourier filtering, and a pure inverse hologram can be extracted. For example, Fig. 2.2.5(b) shows a pattern of Fig. 2.2.5(a) subjected to a two-dimensional Fourier transformation in the ϕ and θ directions. A one-dimensional component of the stripes

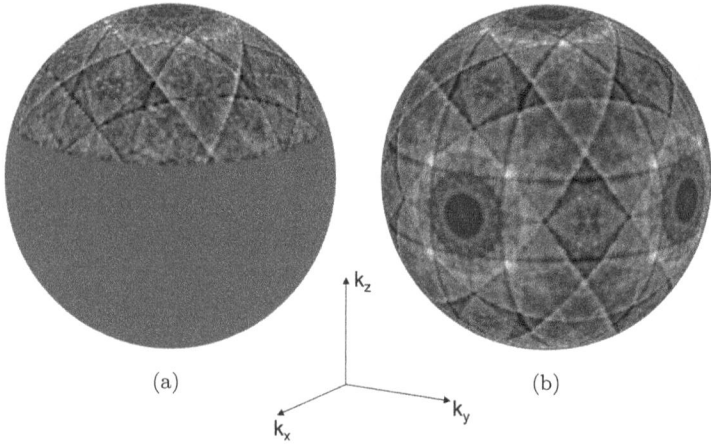

Fig. 2.2.6　Holograms of Au single crystal in k-space. (a) Raw data. (b) Extended hologram using the crystallographic symmetry.

transforms strong spots at $\omega_\theta = 0.0$. Figure 2.2.5(c) shows the pure inverse hologram obtained by applying inverse Fourier-transformation to the pattern in Fig. 2.2.5(b) after removing these spots.

2.2.4.2.　*Extension of hologram using crystallographic symmetry*

The measured hologram pattern is often displayed in a wave vector space (**k**-space) as shown in Fig. 2.2.6, because the Fourier-transform of the hologram in **k**-space directly is just a real space image. We can convert the wavelength of incident X-ray λ, scan angles ϕ and θ to **k**-space (k_x, k_y, k_z) by the following equation:

$$k_x = |k|\cos\phi\sin\theta$$
$$k_y = |k|\sin\phi\sin\theta$$
$$k_z = |k|\cos\theta$$
$$|k| = 2\pi/\lambda \qquad\qquad (2.2.7)$$

Figure 2.2.6(a) shows the hologram pattern after performing the data processing in the previous section. This is the hologram of a gold single crystal measured with incident X-ray of 12 keV ($\lambda = 1.03$ Å). The pattern has only a part of the hemisphere due to the experimental

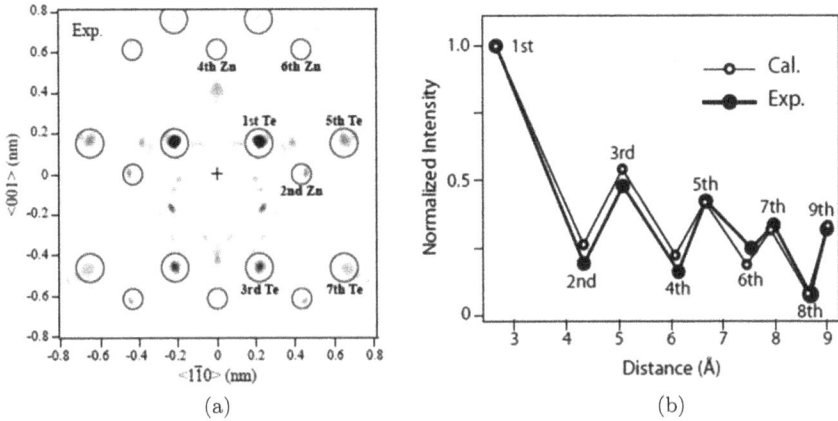

Fig. 2.2.7 (a) Atomic images on (110) plane of ZnTe and (b) the intensity variation as a function of the distance from the origin. This figure is taken from Ref. [16].

limitation of the measurable range of θ, which is typically less than 75°. If Eqs. (2.2.4) or (2.2.5) is applied to this hologram, the resolution of the real space image in z-axis is poorer than those in x- and y-axis, because the k_z range of the hologram is narrower than the other axes.

In order to solve such a problem, a perfect spherical hologram can be obtained using the crystallographic symmetry of the sample. Since this operation uses X-ray standing wave line (Kossel line in the case of normal mode) for determination of the crystallographic orientation, it can be carried out accurately without angular error. The sample used here is a gold single crystal, whose crystal structure is a face-centered cubic lattice. Since the crystallographic symmetry is high, the hemisphere hologram can be extended to a complete spherical hologram as shown in Fig. 2.2.6(b). However, the use of such a symmetry operation is limited to materials with higher crystallographic symmetries.

2.2.5. Relation between Atomic Image and Atomic Fluctuation

Figure 2.2.7(a) shows an atomic image of the (110) plane around Zn atom obtained from a multiple-wavelength hologram of a ZnTe single crystal. The hologram is measured with incident X-rays in the range of 11.0 to 14.0 keV (0.5 keV step). The image intensity is stronger as the atomic number is larger, and it decreases with the increase of the distance from the origin. Figure 2.2.7(b) shows the plot of all the atomic image intensities within a radius of 10 Å. Since Te and Zn atoms alternately appear with the increase of the distance from the origin, the intensity change zigzags. In addition, the intensity change for the calculated hologram was also plotted in Fig. 2.2.7(b). The atomic image intensities of experiments and calculated ones are normalized by the intensities of the strongest first neighbor atoms. As seen from the plots, these variations show good agreement. In this way, it is known that the image intensity can be reproduced from the calculated multiple-wavelength hologram at least in the case of standard samples. Although some readers feel like it is common, the extinction effect of incident X-rays[15] and the twin image problem are more serious when the number of wavelengths is small. Then, the experimental images are hardly reproduced by the calculation for the small number of the wavelengths. By the way, technology to measure hologram with sufficient statistical accuracy contributes to prove the great potential of multiple-wavelength holography.[16] Such a good agreement between the experimental and calculated values shows reliability of X-ray fluorescence holography toward analyses of unknown samples.

In applications of X-ray fluorescence holography, atomic fluctuations have often been estimated for clarifying lattice distortion. Therefore, relationship between deviation of the atom from the ideal position and the change of the atomic image must be investigated. First, we used a simple Zn-Te dimer model to calculate the hologram by giving some distribution to the scatterer Te atom. Here, as shown in Figs. 2.2.8(a) and (b), we assume two types of the distributions, i.e. (a) two-dimensional disk-like distribution in which Te is shifted only in the angular direction with respect to Zn, and (b) three-dimensional spherical distribution. The need for the model in (a) is primarily for first neighbor atom. In general, the bond length is very rigid even changing the surrounded crystal structure. For this reason, the displacement of the first neighbor atom in the angular direction plays an important role to relax the lattice distortion. On the other hand, since it is unnecessary to consider such a constraint for the second neighbor or farther atom, we can use the spherical distribution.

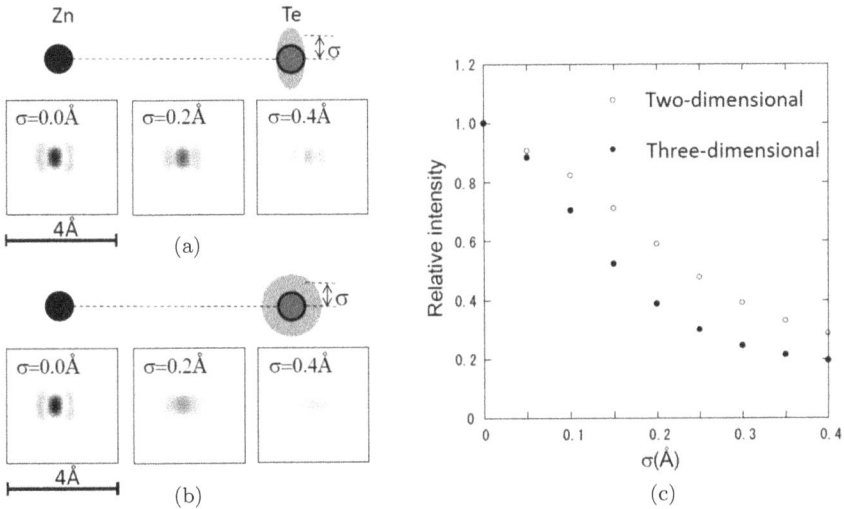

Fig. 2.2.8   Effect of atomic fluctuation on atomic images. Changes of atomic images due to (a) Two- and (b) Three-dimensional fluctuations. (c) Dependence of image intensities. This figure is taken from Ref. [8].

Figures 2.2.8(a) and 2.2.8(b) show the atomic images reconstructed from the multiple-wavelength holograms of Te atoms with the disk-like and spherical Gaussian distributions, respectively. Here, the energy of the incident X-rays were assumed to be 11.0 to 14.0 keV (0.5 keV step). When σ, which is the half width of the half maximum of the Gaussian distribution, increases, the size of the atomic image does not change, because the resolution of the atomic image is about 0.5 Å. While, only the intensity decreases. Therefore, even if distributing atomic positions within this range, it does not influence on the shape of the image. Figure 2.2.8(c) shows the intensity variations as a function of σ. It is known that the intensity change is larger for the spherical distribution than for the disk-like distribution.

2.2.6.   Applications

2.2.6.1.   *Dopants*

Analysis of dopants in single crystals is one of the most important applications of X-ray fluorescence holography. Since 3D atomic images

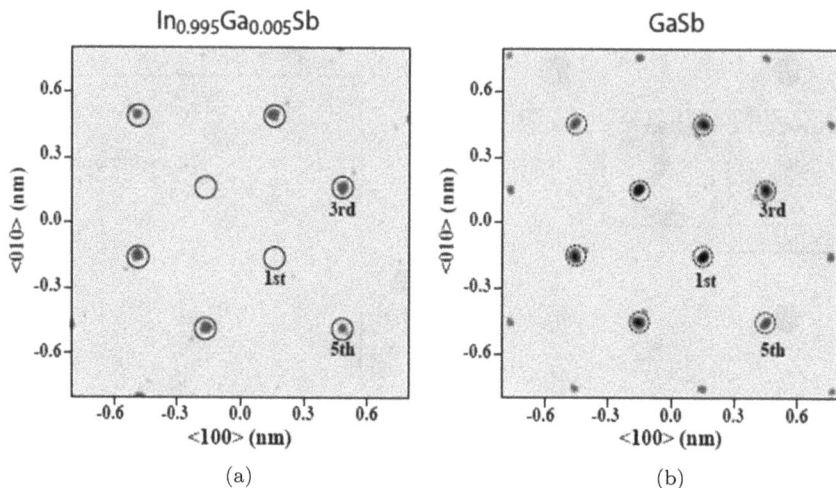

Fig. 2.2.9   Atomic images around dopants. The Sb planes are displayed $a/4$ above Ga, where the a is lattice constant. (a) $In_{0.995}Ga_{0.005}Sb$. (b) GaSb. This figure is taken from Ref. [17].

can be reconstructed over a few nm range, useful information on local lattice distortion around the dopant can be obtained. Such structural properties greatly contribute to understand electronic properties of various materials whose functions are controlled by doping. We measured the X-ray fluorescence hologram of 0.5 at.% Ga-doped InSb ($In_{0.995}Ga_{0.005}Sb$) used for infrared communication device. In addition to this, Ga X-ray fluorescence hologram of GaSb was also observed as a standard sample.[17]

Figures 2.2.9(a) and (b) show atomic images of $In_{0.995}Ga_{0.005}Sb$ and GaSb, respectively. The emitter Ga atoms exist $a/4$ beneath these Sb planes. (a is the lattice constant.) The significant difference between two images is seen at the first neighbor atoms. The first neighbor Sb image is as weak as 37% of that of the GaSb. The reason of this decrement is related to the result of Fig. 2.2.8. Since atomic radii of In and Ga are different, the first neighbor Sb is greatly fluctuated by substituting In sites for Ga. The Ga-Sb bond length at this concentration is 2.67 Å, and the In-Sb bond length is 2.80 Å. Therefore, the first neighbor Sb atom is considered to be shifted inward by 0.13 Å. However, if simply displacing inward, the intensity will not decrease.

The X-ray absorption fine stricture (XAFS) result shows that the Ga-Sb interatomic distance is fairly rigid and the fluctuation in the radial direction is as small as 0.05 Å. Therefore, as shown in Fig. 2.2.8(a), it should be greatly fluctuated only in the angular direction. The estimated distribution in the angular direction is as large as $\sigma_a = 0.37$ Å, which is roughly seven time of $\sigma_r = 0.05$ Å. Such a structural feature was also observed in the structure around Zn in $Cd_{0.04}Zn_{0.96}Te$.[18] Therefore, it can be universal phenomenon that the first neighboring atom around the dopant fluctuates greatly in the angular direction than in the radial direction. The image intensity of $In_{0.995}Ga_{0.005}Sb$ becomes close to that of GaSb with the increase of the distance from the origin. This indicates that the atomic position is stabilized by relaxing the lattice distortion. However, since the image intensity of the third neighbor atom is 71% of that of GaSb, it is known that the lattice distortion is preserved up to this distance.

2.2.6.2. *Disordered systems*

X-ray fluorescence holography is also a powerful technique for disordered systems in which the lattice is largely distorted. For example, a relaxor ferroelectric is one of disordered systems. Since relaxor ferroelectrics show very high dielectric constant and piezoelectricity, they have been industrially used in the world. However, inhomogeneous atomic structures are hardly investigated using conventional methods, and the microscopic origin of their highly functional properties has not been elucidated. In this section, we will introduce the structural evaluation of $Pb(Mg_{1/3}Nb_{2/3})O_3$ studied for a long time.[19]

Figure 2.2.10(a) shows the reconstructed 3D image from the Nb X-ray fluorescence hologram of the $Pb(Mg_{1/3}Nb_{2/3})O_3$ single crystal. The images indicate relative positions with respect to the Nb atom, and are superimposed by all different local environments around the Nb. The cube in Fig. 2.2.10(a) shows one of the eight quadrants of the unit cell. The lead atom should be located at the corner of the cube for the standard perovskite structure. However, the split four images are observed in the <111> direction. This phenomenon can be explained if two kinds of rhombohedral structures, such as acute and obtuse rhombohedrons, which respectively expanded and shrunk in the <111> direction, were

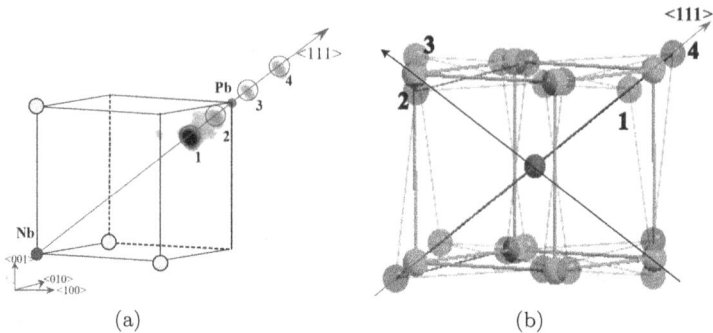

Fig. 2.2.10 (a) Pb atomic images relative to Nb in relaxor ferroelectric $Pb(Mg_{1/3}Nb_{2/3})O3$. (b) Acute and obtuse rhombohedrons derived from (a).

formed as shown in Fig. 2.2.10(b). For example, in the case of the obtuse rhombohedron, when the cube expands in the <111> direction, it shrinks perpendicular to <111>. The former and latter distortions cause outer and inner displacements of the atomic images, respectively, and are reproduced as atomic images 4 and 2, respectively. Similarly, atomic images 1 and 3 can be explained using the obtuse rhombohedron.

Such acute and obtuse rhombohedrons cannot be arranged randomly. However, if they were arranged alternately in the <111> direction and formed a network structure, it is possible to keep a crystal structure entirely. In addition, the network structure is limited to be as short as 1–2 nm. Therefore, the network structure cannot be observed by X-ray diffraction. This result shows that the X-ray fluorescence holography is strong method for the local structural analysis of disordered system.

Acknowledgements

The author, Hayashi, thanks Prof. N. Happo and Prof. S. Hosokawa for their experimental support and analyses.

References

[1] W. Kossel, Z. *Phys.* **94**, 135 (1935).
[2] G. R. Harp, D. K. Saldin, and B. P. Tonner, *Phys. Rev. B* **42**, 9199 (1990).
[3] M. Tegze and G. Feigel, *Nature* **380**, 49 (1996).
[4] M. Tegze, *et al.*, *Phys. Rev. Lett.* **82**, 4847 (1999).
[5] M. Tegze, *et al.*, *Nature* **407**, 38 (2000).

[6] K. Hayashi, *et al.*, *Phys. Rev. B* **63**, R041201 (2001).

[7] S. Marchesini, *et al.*, *Phys. Rev. Lett.* **85**, 4723 (2000).

[8] K. Hayashi, N. Happo, and S. Hosokawa, *J. Electron Spectrosc. Relat. Phenom.* **195**, 337 (2014).

[9] T. Gog, *et al.*, *Phys. Rev. Lett.* **76**, 3132 (1996).

[10] J. J. Barton, *Phys. Rev. Lett.* **61**, 1356 (1988).

[11] J. J. Barton, *Phys. Rev. Lett.* **67**, 3106 (1991).

[12] T. Matsushita, *et al.*, *Phys. Rev. B* **78**, 144111 (2008).

[13] T. Hiort, *et al.*, *Phys. Rev. B* **61**, 830 (2000).

[14] K. Hayashi, *et al.*, *Nucl. Instrum. Methods Phys. Res. A* **467/468**, 1241 (2001).

[15] P. Korecki, *et al.*, *Phys. Rev. B* **69**, 184103 (2004).

[16] N. Happo, K. Hayashi, and S. Hosokawa, *Jpn. J. Appl. Phys.* **49**, 116601 (2010).

[17] S. Hosokawa, *et al.*, *Phys. Rev. B* **87**, 094104 (2013).

[18] N. Happo, *et al.*, *J. Electron Spectrosc. Relat. Phenom.* **181**, 154 (2010).

[19] W. Hu, *et al.*, *Phys. Rev. B* **89**, 140103(R) (2014).

Chapter 3

SURFACE/INTERFACE HOLOGRAPHY

Wolfgang Voegeli
Tokyo Gakugei University

Yusuke Wakabayashi
Osaka University

The most widely used methods for investigating the structure of surfaces are scanning tunneling microscopy (STM) and atomic force microscopy (AFM). They are great tools to examine the topmost atomic layer, because they have basically no sensitivity to the inside of materials. One way to obtain information about the depth direction is to prepare a cross section and observe the structure from the side with scanning transmission electron microscopy, for example. Thinning the sample for this purpose risks to alter the sample, however. In this chapter, we will introduce an electron density analysis method based on surface X-ray diffraction, which gives the three-dimensional structure near surfaces or interfaces without the need for thinning the sample or other elaborate sample preparation methods.

3.1.   X-ray Diffraction from Surfaces

In X-ray diffraction experiments, the intensity of the scattered X-ray wave field is measured as a function of the scattering angle. The amplitude of the scattered wave field is proportional to the Fourier transform of the electron density. The structure could be determined easily by calculating the inverse Fourier transform, if it were possible to directly measure the amplitude (a complex number). Actual experiments observe only the intensity, however, that is the squared absolute value of the amplitude, and the phase information is lost. This problem, common to all diffraction methods, is known as the "phase problem". In the case of single crystal structure analysis by X-ray diffraction, real space structures can be derived by "direct method" software that solves the phase problem. Likewise, the phase problem must be solved for surface X-ray diffraction to determine surface structures. Before discussing this in detail, we will first give a general overview of diffraction from surfaces.

In a crystal, the atoms are arranged periodically with fundamental translation vectors a, b and c. The periodicity gives rise to Bragg reflections as shown in Fig. 3.1. The Bragg reflections are usually denoted

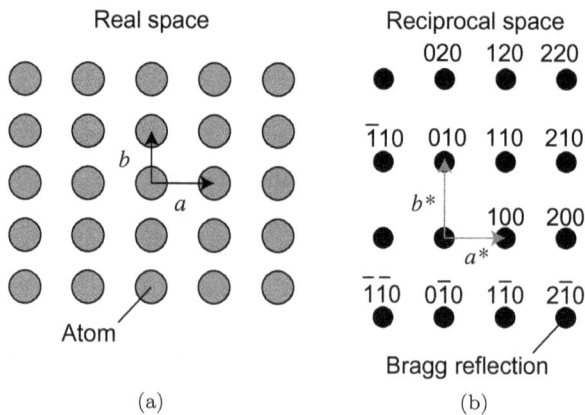

Fig. 3.1   A crystal lattice (a) and the corresponding Bragg reflections (b). For a crystal lattice with basis vectors of the direct lattice a, b and c, Bragg reflections occur in reciprocal space at the positions given by the reciprocal lattice vectors a^*, b^* and c^*. The indices h, k, l of each Bragg reflection are shown by the three numbers next to the reciprocal lattice points. The bar indicates negative numbers.

by the indices h, k and l, which are the Miller indices of the lattice planes that the X-rays are reflected at. The indices can also be thought of as the coordinates of the scattering vector in reciprocal space, which is useful for understanding diffraction processes. The scattering vector is defined as the difference of the wave vectors of the incident and scattered X-rays. It has dimensions of reciprocal length like the wave number, thus it is a vector in wave number space, that is, in the reciprocal space. As indicated in Fig. 3.1 (b), Bragg reflections appear periodically in reciprocal space on the reciprocal lattice with basis vectors a^*, b^* and c^*.[a] The case of a crystal with a three-dimensional periodicity is shown in Fig. 3.2(a), and the corresponding reciprocal space in Fig. 3.2(b), where the positions of reciprocal lattice points are shown by black circles. Figure 3.1(b) is a cross section of this reciprocal space.

We now assume that the crystal has a surface parallel to the c planes.[b] Thus, the translational periodicity is broken in the c-direction at the surface. The periodicity parallel to the surface is unchanged from the interior of the crystal, so the periodicity of the scattered intensity in the a^*- and b^*-directions is the same as that of the infinite crystal. Only the c-direction needs special treatment. Figure 3.3 illustrates how this can be done. In the top row, a one-dimensional lattice in real space with periodicity c and the corresponding lattice in reciprocal space with periodicity c^* are drawn. In the second row, half of the atoms were removed, creating a structure with periodicity $2c$ (gray circles indicate lattice points occupied by atoms, open circles vacant lattice points). In this case, the periodicity of the reciprocal lattice becomes $c^*/2$. As shown in the third and following rows, for a real space periodicity of $4c$, the reciprocal space periodicity is $c^*/4$, for an $8c$ periodicity it is $c^*/8$, and generally for an Nc periodicity it is c^*/N. The limit for $N \to \infty$ gives a macroscopic crystal with a surface. The reciprocal lattice points in the

---

[a]The direction of a^* is perpendicular to the lattice planes spanned by the vectors b and c, its length is the inverse of their interplanar distances. In cubic, orthorhombic and tetragonal crystals, where the lattice vectors are orthogonal, a^* is parallel to a with length $1/a$.

[b]The c plane is the plane spanned by the a and b vectors, or equivalently, the plane normal to the c^* vector.

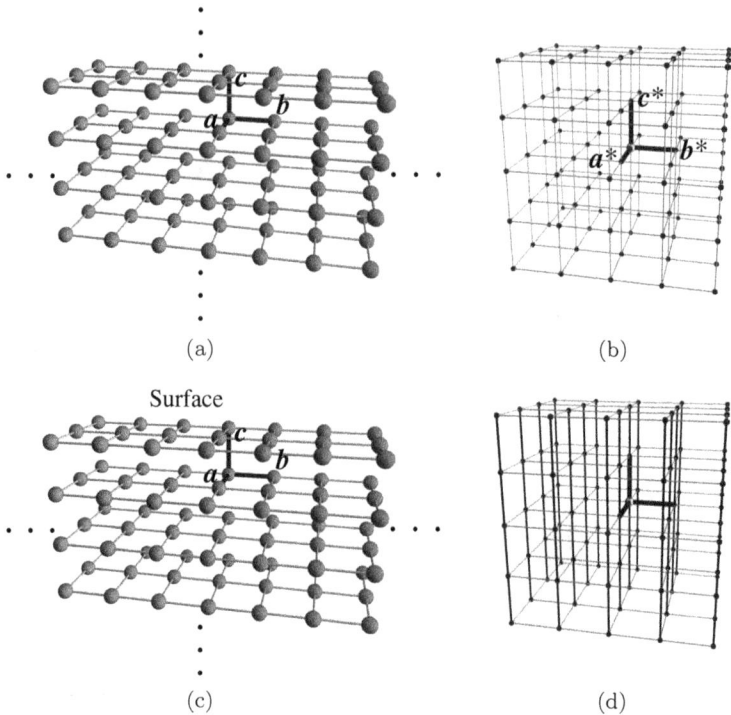

(a)

(b)

Surface

(c)

(d)

Fig. 3.2　Crystal lattice and reciprocal lattice. (a) An infinite three-dimensional crystal lattice. (b) Reciprocal lattice of a crystal with translation symmetry. Bragg reflections appear only at the reciprocal lattice points. (c) Surface of a crystal. The translational symmetry is broken in the c-direction. (d) Scattering from a surface. Rod-shaped scattering appears in the c^*-direction because of the breaking of the translation symmetry.

c^* direction form a continuous line in that case, as shown at the bottom of Fig. 3.3.

For a three-dimensional crystal with a flat surface, similar rod-shaped distributions of the scattered intensity along the c^*-direction appear in reciprocal space (Fig. 3.2(c), (d)). This scattering, caused by the finite extent of the crystal in the c-direction, is called *crystal truncation rod* (CTR) scattering.[1-3]

It should be noted that the scattered X-ray intensity is different for each reciprocal lattice point on the rods. Figure 3.4 shows the intensity distribution calculated for a CTR of a simple cubic lattice. On a linear

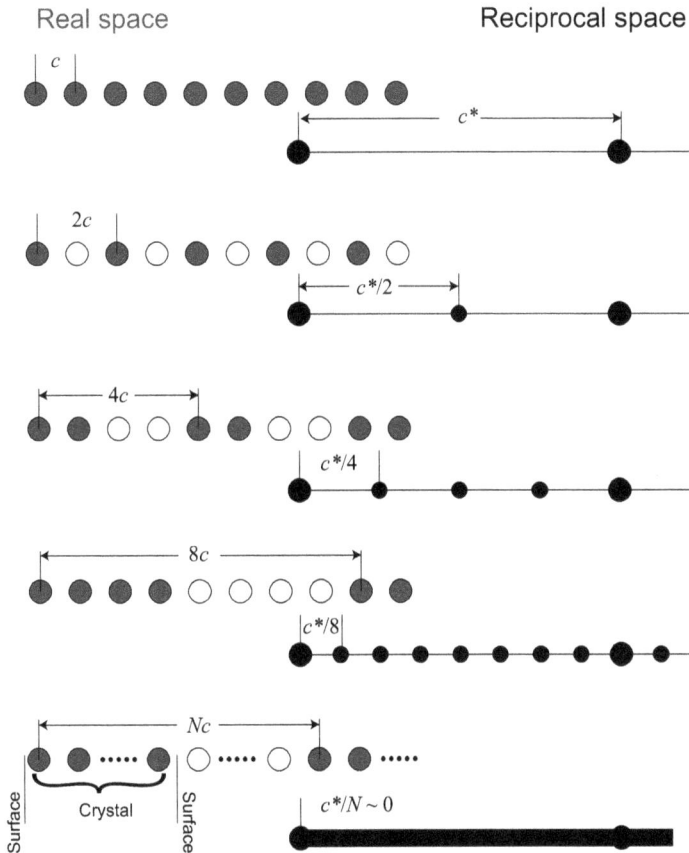

Fig. 3.3 Illustration of the real and reciprocal lattices for various periodicities of a crystal along the c direction. The left side shows the lattice in the real space. Gray and open circles indicate lattice points with a periodicity of c. Gray circles are occupied and open circles are vacant lattice points. From top to bottom, periodicities of c, $2c$, $4c$, $8c$ and Nc are drawn. The black circles on the right side show the corresponding reciprocal lattice points. They have a repeat distance of c^* for a periodicity of c, and a repeat distance of c^*/N for a periodicity of Nc.

scale, only the Bragg reflections from the cubic lattice are visible. The distribution of the CTR scattering can be seen when we plot it on a logarithmic scale. The intensity is equal to the squared amplitude of the Fourier transform of the electron density truncated by a step function at the surface.

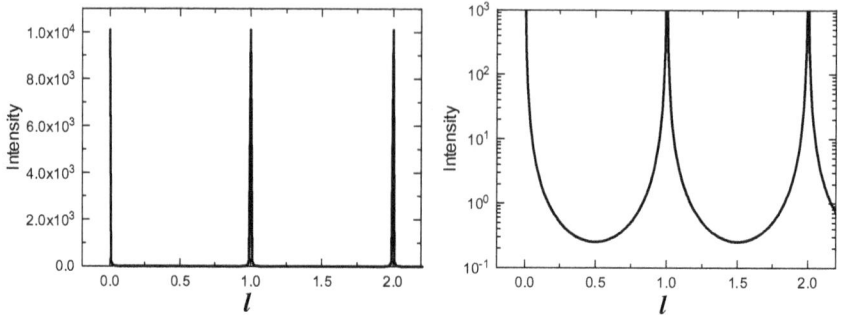

Fig. 3.4 Calculated CTR scattering intensity along the $(00l)$ direction from a flat surface. The intensity axis is linear on the left and logarithmic on the right.

3.2. Scattered Intensity from the Near-Surface Region

The CTR scattering introduced in the previous section is caused by the truncation of the crystal structure at the surface. For real crystals, the surface structure is usually different from the bulk crystal, due to surface relaxation or reconstruction. In addition, an oxidized layer may be present on the surface, or a thin film of a different material may have been grown on it. The coherence length[c] of X-rays from a standard X-ray source is on the order of 1 μm, so the scattering amplitude from near the surface interferes with that from within a depth of about 1 μm inside the sample. We will now assume that there is an imaginary, completely flat surface inside the crystal, as shown in Fig. 3.5. The scattering amplitude from the crystal below the imaginary surface can be calculated from the known crystal structure as shown in the previous section. We will call it $F_{bulk}(Q)$, where Q is the scattering vector. The scattering amplitude from the part of the sample outside of the imaginary surface will be called $F_{surf}(Q)$. The scattering amplitude from the whole sample, $F(Q)$, can then be written as

$$F(Q) = F_{bulk}(Q) + F_{surf}(Q). \tag{3.1}$$

---

[c]The coherence length is a measure for the range in which waves can interfere. The coherence length of synchrotron radiation is about 1 μm, if no special measures are taken to produce coherent X-rays.

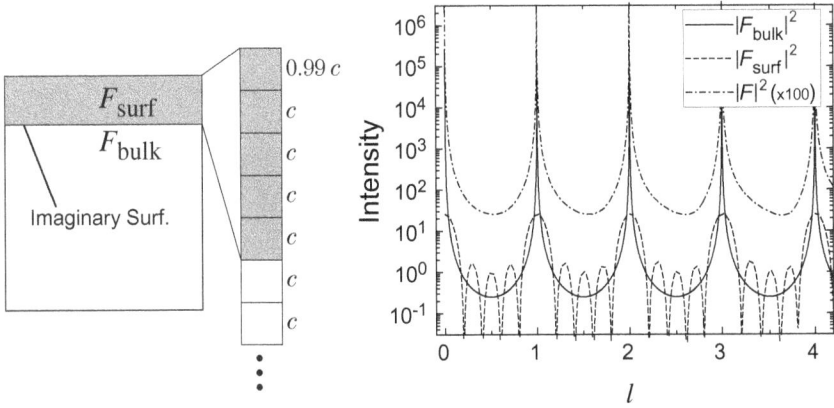

Fig. 3.5   Treatment of the scattering in the case where the surface structure is different from the bulk. The scattering amplitude from inside of an imaginary surface in the crystal (white region), $F_{bulk}(Q)$, and that from outside of the imaginary surface (gray region), $F_{surf}(Q)$, are considered separately. Their interference is observed in experiments. The calculation of the intensity distribution on the ($00l$) axis, shown on the right, was done for a simple cubic lattice with five layers in the gray region. The layer distance of the topmost layer was contracted by 1% with regard to the distance in the bulk.

An example calculation for a crystal with a surface relaxation is shown in Fig. 3.5. In this calculation, the region outside of the imaginary surface is five unit cells thick, and the interplanar distance of the topmost layer is contracted by 1%. The calculation was done for the ($00l$) axis in the reciprocal space of a simple cubic crystal, with equal atomic scattering factors for all atoms. The shift to the right of the minimum of the scattered intensity $|F|^2$ around $l = 3.5$ is caused by the 1% reduction in the interplanar distance of the topmost layer. This shows that CTR scattering is a highly sensitive method, where a structural change in only one atomic layer affects the intensity.

We note that the oscillations of $|F_{surf}|^2$ in Fig. 3.5, due to the finite thickness of the surface layer, disappear when it interferes with the scattering amplitude from inside of the imaginary surface F_{bulk} to give $|F|^2$. The reason is obvious: although we have assigned five layers to F_{surf}, only one layer actually has a structure that is different from the bulk crystal. This is a good example for the difference between adding the X-ray *intensity* scattered from each atom and adding the *amplitude*. If the structure of the five surface layers is really different from the bulk, then the intensity distribution shows oscillations, as can be seen in Fig. 3.8(a). We see from this that the phase of the amplitude carries very important information, notwithstanding that it can not be observed directly. The phase information can be recovered using ideas based on holography. This will be treated in the next section.

3.3. Principle of CTR Scattering Holography

An important feature of surface structure determination is that one can utilize the knowledge of the bulk structure. We can use the phase of $F_{bulk}(Q)$ calculated from Eq. (3.1) to determine the phase of $F_{surf}(Q)$. This is equivalent to the approach taken in holography; $F_{bulk}(Q)$ corresponds to the reference wave, which helps us to derive the information of the object wave, $F_{surf}(Q)$. This idea is called CTR holography.[4,5] Several methods based on this idea have been proposed[6,7] and used for analysis[9-14] since the beginning of the 21st century.

We will here show how to apply holography to CTR scattering following the treatment by Takahashi *et al.*[4] for the case when $F_{surf}(Q)$ is small.[d] The scattered intensity $I(Q)$ can be written as

$$I(Q) = [F_{bulk}(Q) + F_{surf}(Q)]^*[F_{bulk}(Q) + F_{surf}(Q)], \qquad (3.2)$$

where the * indicates the complex conjugate. Omitting the (Q) for clarity, this equation can be rewritten as

$$I = F_{bulk}^* F_{bulk} + F_{bulk}^* F_{surf} + F_{surf}^* F_{bulk} + F_{surf}^* F_{surf}$$

$$\simeq I_0 + F_{bulk}^* F_{surf} + F_{surf}^* F_{bulk} \qquad (3.3)$$

---

[d] $F_{bulk}(Q)$ does not need to be the amplitude of a model where the bulk structure is truncated at a plane, any known reference structure is acceptable. Here, we will assume that a structure model has been constructed for which $F_{surf}(Q)$ is small compared to $F_{bulk}(Q)$.

$I_0 = |F_{bulk}|^2$ is the squared absolute value of the scattering amplitude from the known structure. In the last row, we omitted $|F_{surf}|^2$, the square of a small number. We will consider the following quantity:

$$\frac{I - I_0}{F^*_{bulk}} = F_{surf} + F^*_{surf} \frac{F_{bulk}}{F^*_{bulk}} \tag{3.4}$$

The left side only consists of values that can either be obtained from experiment or can be directly calculated from the structure model. The first term on the right side depends only on the structural change near the surface. The Fourier transform of $\frac{I-I_0}{F^*_{bulk}}$ gives therefore the near-surface structural change together with ghosts originating from the second term on the right side. The ghosts do not necessarily give real numbers, nor do they necessarily give an electron density concentrated in a small region like atoms do, so one can expect that they are not a large obstacle for knowing approximately where the atoms are.[e]

It is easy to do calculations to test the CTR holography method, because the scattered intensity can be calculated when a structure model is assumed. An example calculation from Ref. [4] is presented in the following, for details see that publication. The assumed structure model is a simple cubic lattice with a flat surface, on which one atom per unit cell is adsorbed (Fig. 3.6(a)). $F_{bulk}(Q)$ is obtained from the simple cubic lattice. $F_{surf}(Q)$ corresponds to the scattering from the adsorbed atoms. The adsorbed atoms were placed on the (0.4, 0.6, 0.8) position in fractional coordinates of the unit cell, as shown by the gray circles in Fig. 3.6(a). The CTR scattering intensity distribution along the $(00l)$ rod is shown in Fig. 3.6(b). The calculation according to Eq. (3.4) was done using the (h, k, l) for which $|h|, |k|$ are integers ≤ 10 and $0 < l \leq 10$. The result of the Fourier transform is shown in the square indicated by the dashed line in Fig. 3.6(a). As can be seen, the adsorbed atoms are correctly reconstructed.

---

[e]Ghosts appear in the case of holography using visible light as well, but they usually do not disturb the reconstructed image much.

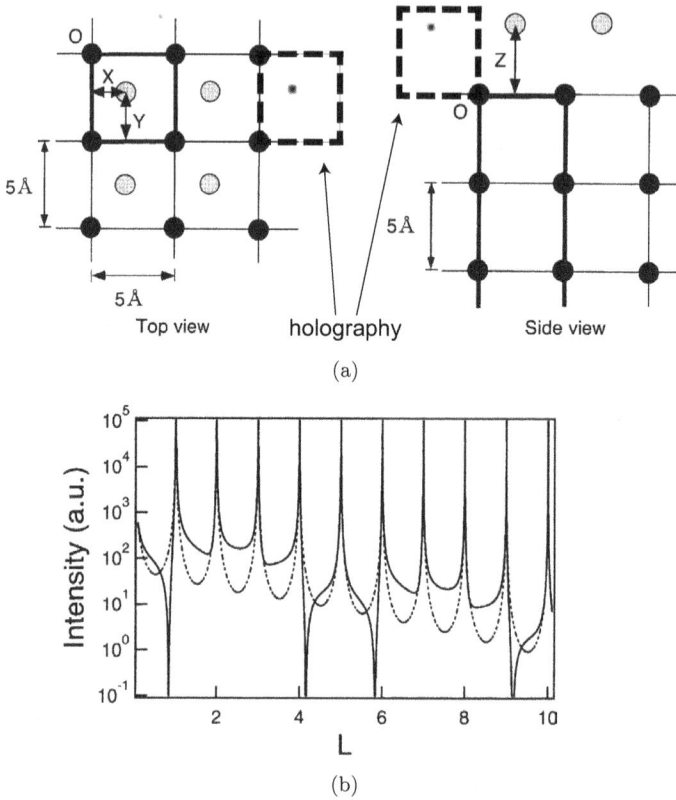

Fig. 3.6   (a) Structure model and result of the holographic reconstruction. (b) The scattered intensity along the (00L) CTR. The dotted line indicates $|F_{bulk}|^2$, and the solid line $|F_{bulk} + F_{surf}|^2$ Figure reproduced from Ref. 4 with permission from Elsevier.

3.4.   Construction of Interface Structure Models Using Holography

An example for CTR scattering holography is the determination of the interface structure of bismuth thin films on silicon.[13] Quantum effects are pronounced in nanostructures of Bi, because of its long Fermi wavelength and mean free path. Bi thin films have been used to investigate the effect of the breaking of the inversion symmetry at surfaces, which is intensively studied in recent years. Bi thin films grown on the Si(111) surface are well suited for this kind of research, because they have a very good crystallinity.

The CTR scattering profile from such a Bi thin film on a Si(111) surface is shown in Fig. 3.7(a). The Bi(001) planes align parallel to the Si(111) planes. The circles indicate experimental data values, while the dashed line has been calculated from an initial model constructed by combining the crystal structures of Si and Bi. The c lattice parameter of Bi, the surface roughness, and the atomic displacement factor were optimized to reproduce the experimental result as closely as possible. The region where the intensity is strong is comparatively well reproduced, but the experimental and calculated values are different in the region where the intensity is weak. This difference was used for the CTR scattering holography analysis introduced in the previous section, with the structure of the Bi film as the reference structure for calculating F_{bulk}. Figure 3.7(b) shows two results using data in a different range along the c^* direction. The only peak that appears in both cases is the one indicated as "wetting layer", showing that atoms exist at this position that are not included in the structure model.

Next, a structure model including atoms at the position of the wetting layer was constructed, and the Coherent Bragg Rod Analysis (COBRA)[6] method,[f] another type of CTR scattering holography, was applied. It gave the electron density profile shown in Fig. 3.7(c). In this way, the interface structure of Bi thin films on Si(111) was determined.

3.5. Complex Surface and Interface Structures

3.5.1. *Ultra-thin films of transition metal oxides*

Transition metal oxides are a group of materials with a wide variety of properties, for example ferroelectricity, metallicity, ferromagnetism, antiferromagnetism, or superconductivity. There are several classes of crystal structures, such as perovskite or spinel structures, which results in many materials having different properties with similar structures. It is possible to create interfaces between two materials with different

---

[f]This method is influenced more strongly by the initial structure model than the method explained above, but has the advantage that it does not produce ghosts. It is most useful for refining a model after the structure has been understood to some extent.

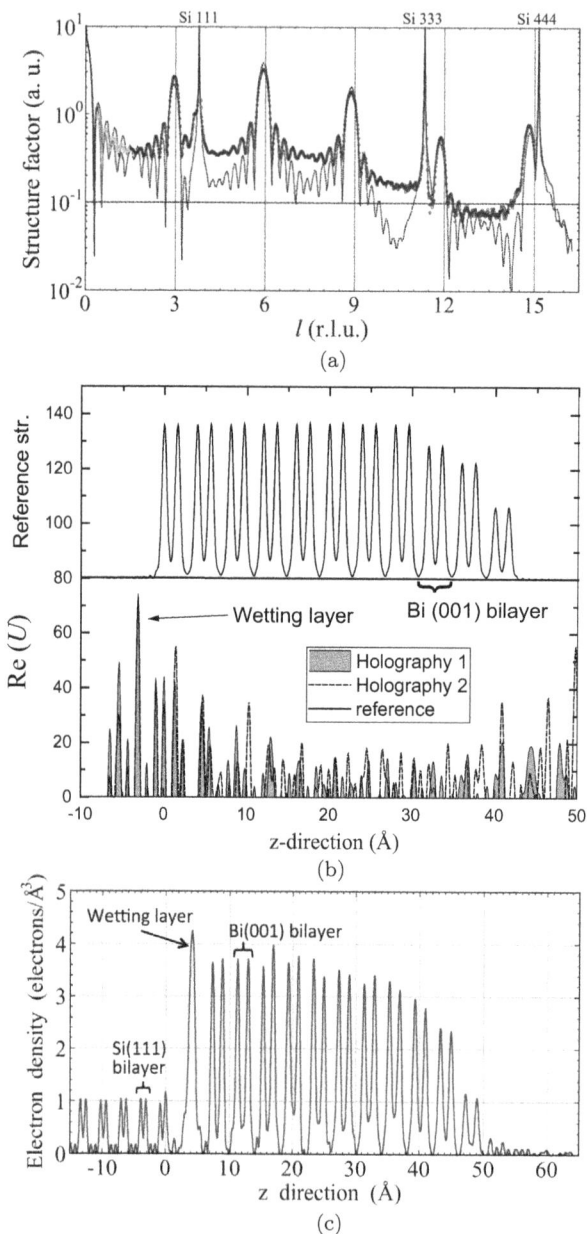

Fig. 3.7   (a) CTR scattering profile from a Bi thin film on Si(111). (b) Result of the CTR scattering holography analysis. (c) Result of the electron density analysis based on the improved structure model. Figure reproduced from Ref. 13. Copyright (2011) by the American Physical Society.

properties, for example ferromagnetic and antiferromagnetic, without disturbing the periodicity of each material. Such interfaces are studied widely, because of their interest for both basic physics and applications. Especially attracting attention is that the interface between two materials can have properties that are different from both of them. A good example is the formation of a highly conductive electron gas at the interface between the two band insulators $LaAlO_3$ and $SrTiO_3$.[15] This interface has been extensively studied; ferromagnetism and superconductivity have been reported as well. We will here focus on the interface structure.

The most widely used method for characterizing the films after growth is cross-sectional scanning transmission electron microscopy (STEM), in which STEM images are taken on sliced specimens. Although this method provides atomic resolution images, the resolution is not sufficient to observe the atomic displacements related to the electric polarization. Figure 3.8 shows an example of analyzing X-ray CTR scattering with the COBRA method, introduced in the previous section, to obtain the electron density with high resolution.[9] Atomic displacements can be obtained from the figure with an accuracy that is significantly higher than the atomic displacements caused by the polarization inside the dielectric (about 0.1 Å). These atomic displacements can be used for comparing to theoretical calculations, and for theoretical interpretations based on the measured structure.[9, 11, 12]

3.5.2. *Organic semiconductors*

Electronic devices using organic semiconductors are already commercially available, a prominent example is light-emitting devices (organic electroluminescence displays). Their physical understanding is still insufficient, however; the mechanism for electric conduction in organic semiconductors is controversial, for example. In field-effect transistors made from organic semiconductors, the electric conduction is limited to a thickness of about one molecular layer at the interface. It is therefore of prime importance to understand the interfaces. In bulk organic materials, electric conduction is dominated by the hopping integral between molecules, which has a strong dependence on the structure. Knowledge

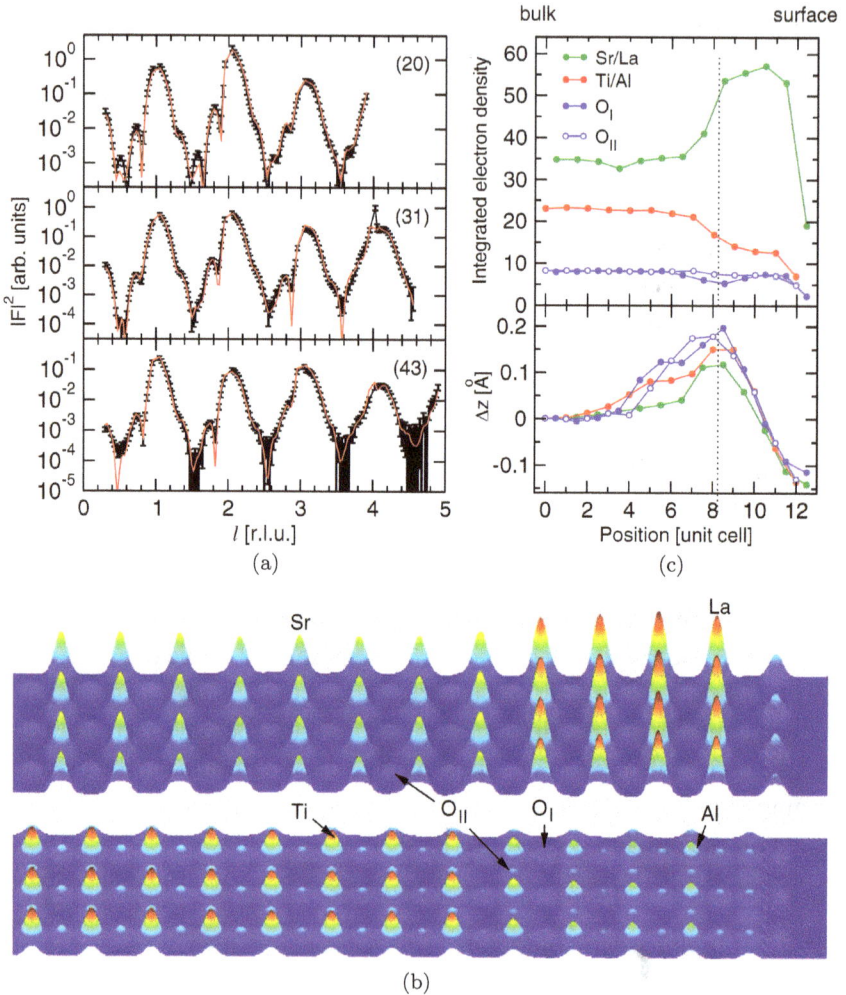

Fig. 3.8 Results of the CTR scattering experiment of a LaAlO$_3$/SrTiO$_3$ interface. (a) CTR scattering profile, (b) electron density, and (c) number of electrons and atomic displacements for each atomic site obtained from the electron. Figure reproduced from Ref. 9. Copyright (2007) by the American Physical Society.

about the surface structure is therefore necessary for understanding the physical processes involved.

The structure of molecular solids is much more complicated than silicon or the simple cubic lattices that we have used as examples till

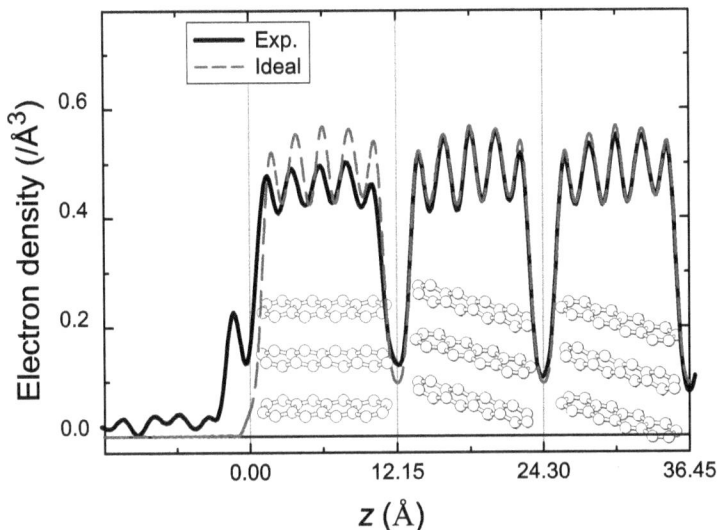

Fig. 3.9 Electron density profile of tetracene single crystals obtained from CTR scattering holography (solid line). The dashed line indicates the expected electron density if there were no surface relaxations. It was found that the molecules in the topmost layer are rotated with regard to deeper layers. The surface is at $z = 0$. The peak at $z < 0$ is caused by adsorbed material on the surface. Figure reproduced from Ref. 14. Copyright (2014) by Springer Nature.

now. This poses the question whether it is really possible to characterize them with CTR scattering holography. Such experiments have been reported for the first time for rubrene and tetracene single crystals.[10,14] Plate-shaped crystals with large c-plane facets of these materials can be grown. AFM measurements show that the surface consists of flat terraces more than 1 μm wide. The results for tetracene of an electron density analysis using COBRA based on the $(00l)$-axis CTR scattering profile are shown in Fig. 3.9. It was found that there is a layer of adsorbed material on the surface and that the positions of the molecules in the topmost layer are relaxed from the bulk positions. The surface relaxations are only evident in the topmost layer, in contrast to metals, where they extend to several layers below the surface. This can be understood as a consequence of the strong localization of the electrons on the molecular orbitals and the weak bonding between the molecules.

3.6.   Features and Limitations of CTR Scattering Holography

We have seen in the previous sections that CTR scattering holography is a powerful characterization method. Its advantages are:

- X-rays are not influenced by electric or magnetic fields.
- Non-contact measurement.
- High flexibility for sample environments, because of the high penetration depth of X-rays.
- High resolution, sufficient to observe polarization effects.
- Structure measurements are not influenced by the electronic structure.

Surface observation methods using electrons have a small penetration depth and often have difficulties with non-conducting samples. CTR scattering, on the other hand, does not suffer from these problems. It has some limitations, however:

- The surface must be flat on the atomic scale in order to obtain any signal.
- Elements with close atomic numbers are difficult to distinguish.
- Synchrotron radiation is needed as the signal intensity is weak.
- It is insensitive to the details of the electronic structure because the total electron density is measured.

The last "limitation" is of course the flip-side to the advantage that the structure can be determined independent of the electronic structure. A final point is that the CTR scattering holography method presented here can only give information that has the same in-plane periodicity as the substrate, because it uses the interference between the scattering amplitudes from the surface and the substrate.[g] Scattering from surface superstructures can not be used for analysis, because there is no corresponding scattering from the substrate that could serve as the reference

---

[g]This is only a limitation of CTR scattering holography. Structures of surfaces that have a different periodicity to the substrate can be analyzed using least-squares fitting of a structure model to the experimental intensities. Many surface structures have been determined in this way. We will not discuss this further, since it is out of the scope of this chapter.

wave. Structures that have a different periodicity in the surface plane are not completely invisible, however, their electron density projected onto the surface normal axis contributes to the scattered intensity of the $(00l)$ rod. The adsorbate in Fig. 3.9 is an example for this. CTR scattering can be used to acquire information that is difficult to obtain using other methods, if the points mentioned above are kept in mind.

References

[1] S. R. Andrews and R. A. Cowley, *J. Phys. C* **18**, 6427– (1985).
[2] I. K. Robinson, *Phys. Rev. B* **33**, 3830 (1986).
[3] T. Takahashi, *et al.*, *Surf. Sci.* **191**, L825 (1987).
[4] T. Takahashi, K. Sumitani, and S. Kusano, *Surf. Sci.* **493**, 36 (2001).
[5] D. Saldin, *et al.*, *Computer Physics Communications* **137**, 12 (2001).
[6] M. Sowwan, *et al.*, *Phys. Rev. B* **66**, 205311 (2002).
[7] R. Fung, *et al.*, *Acta Cryst. A* **63**, 239 (2007).
[8] D. K. Saldin and V. L. Shneerson, *J. Phys.: Condens. Matter* **20**, 304208 (2008).
[9] P. R. Willmott, *et al.*, *Phys. Rev. Lett.* **99**, 155502 (2007).
[10] Y. Wakabayashi, J. Takeya, and T. Kimura, *Phys. Rev. Lett.* **104**, 066103 (2010).
[11] S. A. Pauli, *et al.*, *Phys. Rev. Lett.* **106**, 036101 (2011).
[12] R. Yamamoto, *et al.*, *Phys. Rev. Lett.* **107**, 036104 (2011).
[13] T. Shirasawa, *et al.*, *Phys. Rev. B* **84**, 075411 (2011).
[14] H. Morisaki, *et al.*, *Nature Commun.* **5**, 5400 (2014).
[15] A. Ohtomo and H. Y. Hwang, *Nature* **427**, 423 (2004).

Chapter 4

IMAGING OF NANOSTRUCTURES AND SINGLE MOLECULES

Subchapter 4.1

DIFFRACTIVE IMAGING

Kazutoshi Gohara
Hokkaido University

Hiroyuki Shioya
Muroran Institute of Technology

Jun Yamasaki
Osaka University

Diffractive imaging has been studied in the fields of visible light, X-rays, and electrons. In this section, examples of experiments and the foundation of phase retrieval are presented. In 4.1.1 and 4.1.2, the basics of transmission electron microscopes and examples of diffractive imaging in electron microscopes are introduced. In 4.1.3, phase retrieval is presented from a mathematical viewpoint.

4.1.1. Basics of transmission electron microscopes

4.1.1.1. *Mechanism of transmission electron microscopes*

In electron microscopes, atoms and various nanostructures can be imaged by using beams of electrons, which have wavelengths much shorter than that of visible light. Transmission electron microscopy (TEM) is a conventional method whereby magnified images of samples are formed by electron beams after transmission through the sample. Note that the abbreviation "TEM" is also used for microscopes based on TEM imaging. In TEM, electrons emitted from a source are accelerated by a high voltage (several tens of kV to ~3000 kV) and impinge upon a sample. The high-energy electrons that are transmitted through the sample are bent by magnetic lenses to form a magnified image. Among the many lenses incorporated in a TEM, the objective lens located just below the sample plays a major role in producing the magnified image. Figure 4.1.1 shows a schematic illustration of the optics around an objective lens.

When the incident electrons pass through the sample, a fraction of them is scattered. For a crystalline sample, the electrons undergo Bragg diffraction (see Chapter 3). Figure 4.1.1 shows G and –G diffracted waves as representative of Bragg diffraction and whose wave direction differs from that of the incident beam. Each diffracted beam and nonscattered

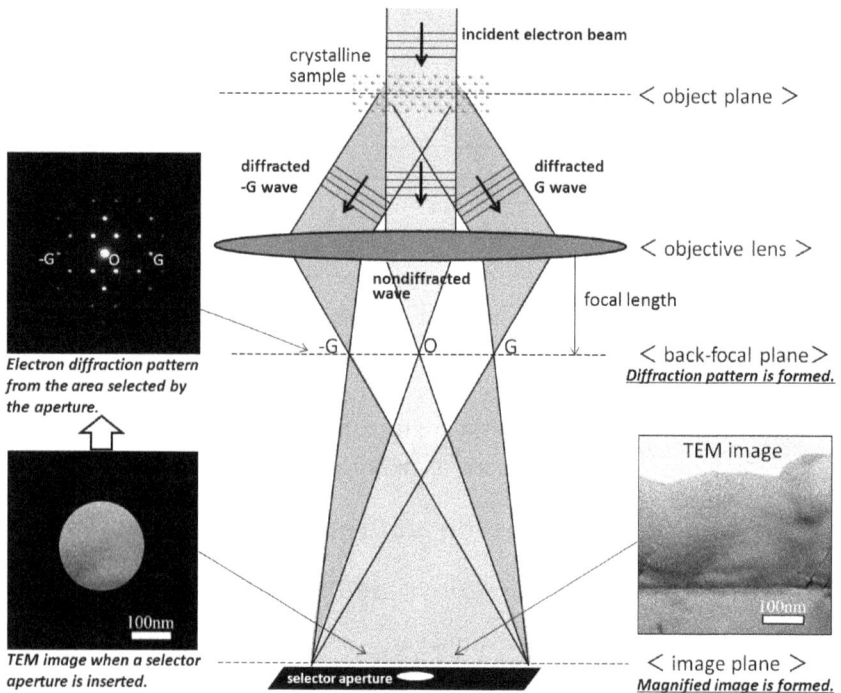

Fig. 4.1.1 Formation of a magnified image and corresponding electron diffraction pattern of a sample formed by objective lens in a TEM.

beam converges on the focal plane of the objective lens (i.e., the back-focal plane). As seen in Fig. 4.1.1, the positions of the convergence points depend on how the beam is deflected by the sample, which means that the angular distribution of the scattered waves appears on the back-focal plane. The squared modulus of the wave field on the back-focal plane is equivalent to the Fraunhofer diffraction pattern formed at infinite distance. As explained in Chapter 3, the diffraction pattern corresponds to the modulus squared of the Fourier transform of the sample structure. To be precise, X-ray and electron diffraction patterns correspond respectively to the Fourier transform of the electron-density distribution and of the potential distribution for electrons (mainly the Coulomb potential as formed by the constituent atoms).

Downstream from the back-focal plane, the converged beams gradually expand and overlap each other at the image plane of the objective lens, where the magnified image of the sample is formed. Note that actual TEMs have some lenses also below the image plane. These lenses project the sample image in the image plane onto a detector and provide additional magnification. The detector then records the magnified TEM image. Since lenses in TEMs consist of electromagnets, their power is easily changed by modifying the coil currents, which is not possible in optical microscopes. Therefore, users can easily change various imaging parameters such as magnification and illumination intensity as well as focusing on the sample. Inserting a plate with a pinhole (called a "selector aperture") in the image plane (see bottom of Fig. 4.1.1) produces an image on the detector of the sample selected by the aperture (see lower-left side in Fig. 4.1.1). By adjusting the power of the lens below the image plane, the back-focal plane can be focused onto the detector (instead of the image plane). Thus, we can observe the diffraction pattern from the area selected by the aperture, as shown on the left side of Fig. 4.1.1. This technique is called selected-area diffraction and is the most common technique used to observe diffraction patterns with TEMs.

4.1.1.2. *Characteristics of image formation in TEM*

In TEMs, incident electrons have kinetic energies over several tens of keV, so they are hardly absorbed by the sample in ordinary TEM observations, where the sample thickness is typically less than 1 μm.

Except for backscattered electrons, most incident electrons pass through the sample, which means that the brightness of the TEM image remains relatively constant regardless of whether a sample is present. In other words, the sample should be invisible in TEM images because of a lack of image contrast. However, in actual TEM images, image contrast arises because of the objective aperture inserted at the back-focal plane (and because of other apertures positioned near the optical axis). Since these apertures block electrons that are scattered over angles greater than that accepted by the aperture radii, matter appears darker than the vacuum because the former scatters whereas the latter does not. Therefore, a thick sample area and/or a sample area containing heavy elements appear darker than other areas in TEM images. This is called "mass-thickness contrast." In the same manner, blocking beams that are Bragg reflected by crystalline samples also results in intensity modulations (called "diffraction contrast"). These are the main methods of obtaining image contrast in medium-resolution TEM images and are applicable provided that the sample is in focus. When the sample is out of focus, the scattered and nonscattered beams do not perfectly overlap at the image plane. Because this introduces a complicated image contrast that does not correspond to the sample structure, the focus of the objective lens must be carefully adjusted. Quantum theory tells us that the TEM-image intensity is proportional to the modulus squared of the electron wave function at the detector. In other words, the image intensity in medium-resolution TEM reflects only the amplitude information and does not include the phase information of the complex wave function transmitted through the sample.

Unlike the situation for medium-resolution TEM images, intentional slight defocusing (i.e., modulation starting from an in-focus condition) instead of blocking by apertures is used to obtain image contrast in high-resolution TEM (HRTEM). As explained in the next section, the phase of an incident electron wave advances more rapidly in a material than in vacuum because of the difference in potential energies. Because a potential well formed by a single atom advances the phase only a small amount in most cases, the wave function after transmission through an extremely thin material such as graphene shows only a slight phase modulation at the atomic positions (this is called the weak-phase object approximation[1]). In the case of crystalline materials thicker than several nm, many atoms lie on the path of the incident electrons (called atomic columns). As the result, in the wave function after the sample, both the amplitude and phase are significantly modulated at the column positions. Therefore, by imaging the modulus squared of the wave function, such atomic column positions are observed. Unfortunately, forming a HRTEM image that shows the correct atomic positions is not straightforward because of imperfections in the objective lens (lens aberrations). Intentional defocusing from the in-focus condition is known as an effective way to minimize the influence of the aberrations and to maximize image contrast (this is called Scherzer defocusing[1]). Thus not the in-focus condition but Scherzer defocusing condition gives the best spatial resolution of HRTEM, which is limited by lens aberrations in the objective lens.[1]

4.1.2. Examples of diffractive imaging in electron microscopes

As mentioned above, diffraction patterns are based on the Fourier transform of the sample structure. In the case of crystalline materials, a cross section of the reciprocal lattice appears as the diffraction pattern under the kinematical approximation (see Chapter 3). To be precise, the diffraction intensity is the modulus squared of the diffracted wave, which is the Fourier transform of the real-space wave just below the sample. Based on this relationship, the real-space wave can be reconstructed numerically from the diffraction pattern. This method is called "diffractive imaging." As mentioned in the previous section, the following problems are posed by imaging with lenses: (1) The phase distribution of a complex wave function is not observed. (2) HRTEM to observe atoms is severely affected by lens aberrations. As explained below, diffractive imaging can solve these problems.

4.1.2.1. *Phase imaging and observations of nano-electromagnetic fields*

Based on the Fourier transform relationship, the diffraction intensity, which is the modulus squared of the diffraction wave, should be consistent with the amplitude and phase of the real-space wave. This consistency works as a constraint to reconstruct the real-space wave from the diffraction pattern. Thus, we can obtain the phase image, which is invisible when imaging with lenses, as mentioned above. The details of the reconstruction procedures are described in Section 4.1.1.3. This section introduces some examples of phase imaging.

The average potential energy in materials due to constituent atoms is lower for incident electrons than that in the vacuum, which means that incident electrons are accelerated inside materials. In other words, the average wavelength of the electrons shortens. Therefore, the phase of the electron wave passing through a material advances more rapidly than when traveling the same length in vacuum. The phase difference is known to be approximately proportional to the difference in potential energies. Figure 4.1.2 shows reconstruction of the phase image of a wave that passed through a wedge-shaped silicon crystal, as shown schematically in Fig. 4.1.2(a).[2] Figure 4.1.2(b) shows a TEM image of the wedge and selector aperture. The corresponding diffraction pattern

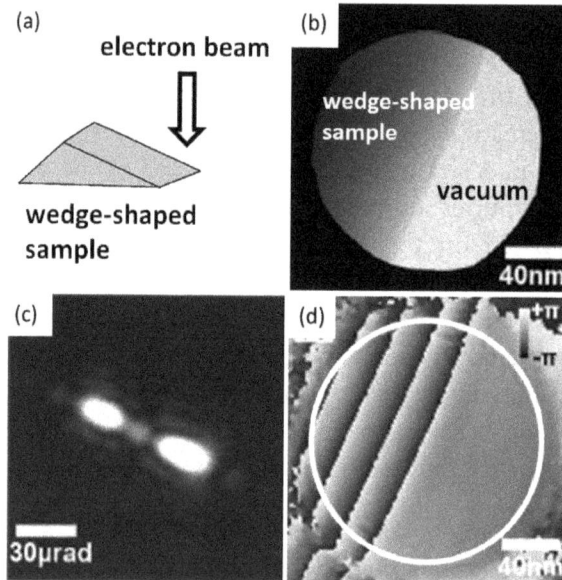

Fig. 4.1.2   Phase imaging of wedge-shaped sample.[2] (a) Schematic diagram showing the incident beam direction. (b) TEM image of sample and selector aperture. (c) Selected-area diffraction pattern. (d) Phase image reconstructed from the data shown in panels (b) and (c).

appears in Fig. 4.1.2(c). The reconstructed phase image in Fig. 4.1.2(d) shows that the phase advances with distance from the sample edge. This result reflects the fact that the phase advance is proportional to the material thickness; that is, to the distance traveled in the material.

In the same manner, electric fields existing inside materials and extending into the vacuum may be visualized through phase imaging. This is possible because the electrostatic potential V generates the potential energy $(-eV)$ for electrons. Figure 4.1.3 shows a visualization of a nano-electric field obtained by the phase imaging. As shown by the arrows in Fig. 4.1.3(b), the electric-field vectors, which are given by the negative gradient of the electrostatic potential, are perpendicular to the equi-phase lines. In addition, nanomagnetic fields may be visualized by phase imaging of potential energies from Lorentz forces.

To summarize, phase imaging based on electron diffractive imaging allows nano-electromagnetic fields to be visualized in and around samples in addition to detecting sample configurations such as thickness and composition. This method is effective for observing various functional materials with nanometer-sized electromagnetic features, such as semi-conductor devices with efficiently designed dopant distributions and insulating or magnetic multilayers.

4.1.2.2. Observations of atomic arrangements I

As shown in Fig. 4.1.2, the spatial information extending over 100 nm appears in a small-angle-scattering pattern (typically in angular regions

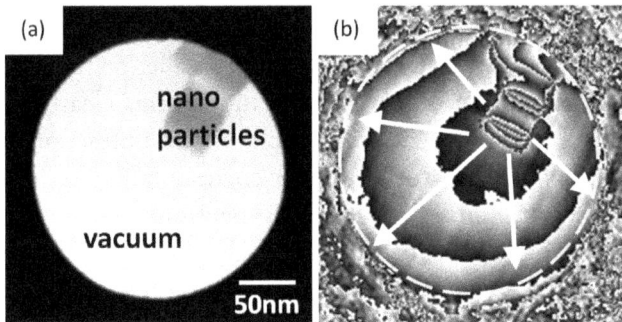

Fig. 4.1.3 Observation of charged nanoparticles. (a) TEM image. (b) Reconstructed phase image showing electric field around particles.

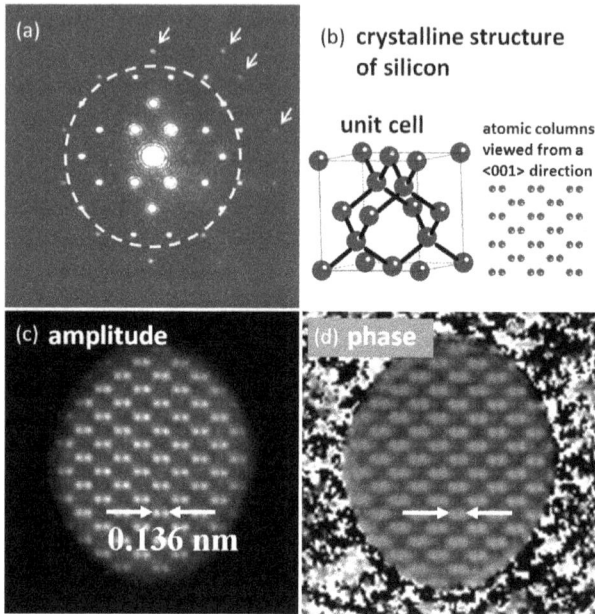

Fig. 4.1.4 High-resolution reconstruction of atomic columns in silicon crystal.[3] (a) <011> diffraction pattern. (b) Schematic of crystalline structure of silicon. (c) Amplitude and (d) phase of the reconstructed wave.

less than 1 mrad for 200 kV incident electrons). However, spatial information at the atomic level appears as Bragg reflections in higher angular regions (typically several tens of mrad for 200 kV incident electrons). This makes sense if we recall that fine spatial information is Fourier transformed to distant positions in reciprocal space. Figure 4.1.4 shows atomic columns reconstructed from the diffraction pattern from crystalline silicon with the electron beam parallel to the <011> axis.[3] The atomic column separation of 0.136 nm appears clearly in both the amplitude and phase images. The circle in the diffraction pattern in Fig. 4.1.4(a) shows the spatial frequency corresponding to the resolution restriction in typical HRTEM images, as mentioned previously. In other words, HRTEM images consist of the Fourier information only inside the circle. Conversely, diffractive imaging based on the entire diffraction pattern should naturally include finer structural information from outside the circle, leading to a higher spatial resolution than for imaging by lenses.[4]

4.1.2.3. *Observations of atomic arrangements II*

Low-acceleration voltage in electron microscopes is essential for reducing electron beam damage in observations of nanostructure materials such as organic molecules consisting of light elements of carbon atoms. However, it is generally difficult to realize atomic-resolved imaging using low-acceleration voltage due to lens aberrations. Therefore, diffractive imaging is expected for high-resolution imaging using low-acceleration voltage. In this section, diffractive imaging is presented as an example of atomic-resolved imaging of a carbon nanotube using a scanning electron microscope (SEM) with low-acceleration voltage of 10–30 kV.[5,6]

The intensity of the diffraction pattern of a single-wall carbon nanotube (SWCNT) with a diameter of a few nanometers becomes broad due to oscillation and bending. Therefore, a SWCNT that is fixed at opposite ends was sought out. Figure 4.1.5 shows an example of SEM image of a specimen and a diffraction pattern in the inset obtained at an acceleration voltage of 30 kV, beam divergence of 0.15 mrad, and exposure time of 30 s. The diffraction pattern showed that the diameter

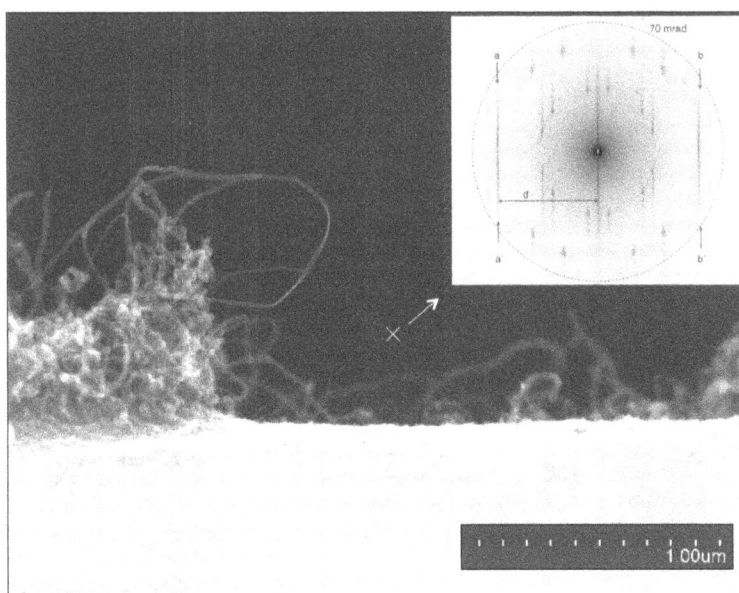

Fig. 4.1.5   SEM image and diffraction pattern (inset) of SWCNT.

of the SWCNT was 3.2 nm and that the SWCNT was tilted at an angle of 8 ± 2 degrees to a horizontal direction.

Before phase retrieval was applied, inelastic scattering around the direct beam and a uniform background were reduced from raw data. In addition, deconvolution was executed during phase retrieval in order to correct the intensity convolved with the divergence of the incident beam.[7]

The image in Fig. 4.1.6(a) was obtained as a result of reconstruction by the phase retrieval algorithm (see the section 4.1.3). (b) shows the corresponding multi-slice simulated image. (c) and (d) are enlarged views of distinctive parts of (a) and (b), respectively, and (e) shows the corresponding atomic arrangement. Comparison of these images revealed that the retrieved image agrees with the atomic arrangement of a SWCNT with a diameter of 3.2 nm.

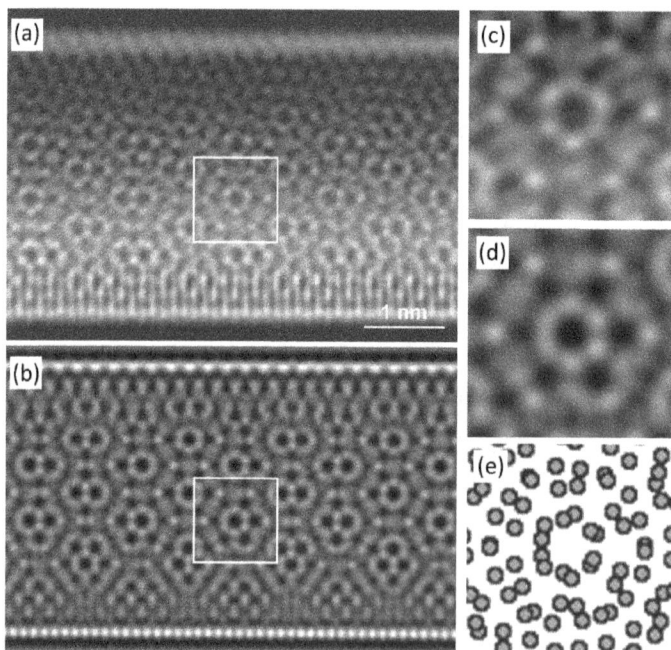

Fig. 4.1.6   Atomic-resolved electron diffractive imaging of SWCNT (a) Reconstructed image, (b) Simulated image. (c) Enlargement of squared region in (a), (d) Enlargement of squared region in (b), (e) Model of atomic arrangement.

4.1.3. Phase retrieval

Phase problems arise while measuring the scattering of object waves. The Fourier intensity can be observed by detecting the diffraction wave, but the phase is lost. Thus the problem is related to the wide-range field, from the atomic to astronomic scales. In this section, the fundamental review of this problem is presented in the terms of the lost-phase estimation using the prior information.

4.1.3.1. *What is phase retrieval?*

We start with formulating the foundation of the retrieval. Let $f(x)$ be an unknown complex-value function in the object space X, which satisfies $\int_X |f(x)|dx < \infty$. The objective is to find $f(x)$. Using the function $K(x, y)$ on $X \times Y$, the transformation of the functions on X to that on Y is expressed as follows.

$$g(y) = \int_X K(x, y)f(x)dx \qquad (4.1.1)$$

The function K is called the kernel function, and Fourier transform is a typical kind. In here, $\bar{K}$ is also defined on $X \times Y$, and satisfying the following.

$$f(x) = \int_Y \bar{K}(x, y)g(y)dy \qquad (4.1.2)$$

The relationship between two kernels is related to the Fourier transform and its inverse with a relevant constraint. f and g are equivalent in the meaning of the transform. If the complete observation of g is not established, the reconstruction of $f(x)$ becomes to be difficult. For example, involving the noise δ_{noise} with zero mean and finite variance, the observed space Y is presented as

$$g_{obs}(y) = g(y) + \delta_{noise}(y) \qquad (4.1.3)$$

By the transform $\bar{K}$, the noise-influenced f_{noise} can be obtained as follows.

$$f_{noise}(x) = \int_Y \bar{K}(x, y)\delta_{noise}(y)dy \qquad (4.1.4)$$

The difference between target f and f_{noise} is given by the inverse transform of the noise factor in the following.

$$|f(x) - f_{\text{noise}}(x)| = \left| \int_Y \bar{K}(x,y)\delta_{\text{noise}}(y)dy \right| \qquad (4.1.5)$$

Using the statistical property of the noise, the estimation $f(x)$ is calculated. The restricted information of the target is a type of inverse problem, but the setting of the phase problem is more difficult than the case of the noise contamination. $|g(y)|^2$ is obtained as an observation of g, i.e., to reconstruct f without $\Phi(y)$ where $g(y) = |g(y)|e^{i\Phi(y)}$.

Let us define the Fourier phase retrieval. The function including the target is called the object, and the related space is called the object domain. The object is a finite-volume function, and its transform is a complex-valued function in the Fourier domain. $\rho(r)$ is an object function on the space X which is a two-dimensional discrete space $N \times N$. The discrete Fourier transform of ρ is expressed as follows.

$$F(k) = \sum_{r \in X} \rho(r)e^{-i2\pi k \cdot r/N} \qquad (4.1.6)$$

where $F = |F|e^{i\Phi}$ on the space Y which is equal to $N \times N$, and its phase Φ is lost while measuring the Fourier intensity. If $\hat{\Phi}$ is an estimator of the phase, the reconstructed object is given by

$$\rho(r) = \frac{1}{N^2} \sum_{k \in K} F'(k)e^{i2\pi k \cdot r/N} \qquad (4.1.7)$$

where $F' = F_{\text{obs}}e^{i\hat{\Phi}}$ and F_{obs} is the observation that excludes the phase. Some the constraints of the object or Fourier domains are required for reconstructing the phase. The object-domain constraint is related to prior information concerning the target object, and the Fourier-domain constraint is related to the observed measurement. The problem requires finding a function satisfying both domain constraints.

4.1.3.2.   *Phase retrieval algorithm*

The reconstruction of the lost phase in the intensity measurements requires the procedure that finds an object satisfying the both the object-domain and Fourier-domain constraints. However, the procedure

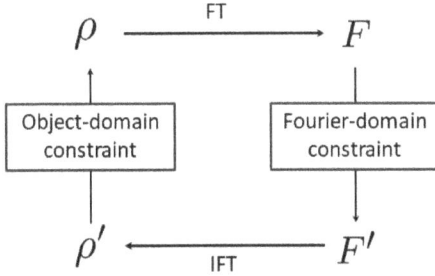

Fig. 4.1.7   GS-diagram with the iterative Fourier transform and constraints.

is based on the respective usage of each constraint because the both constraints are defined in two different domains.

The algorithm of phase retrieval was first presented by Gerchberg and Saxton.[10] An object is a complex-valued function, with the target and Fourier intensities given. A schematic of the algorithm is shown in Fig. 4.1.7 with the iterative procedure comprising Fourier and its inverse transforms. At the beginning of the algorithm, any object function ρ is settled as an initial. Then F is obtained by the Fourier transform of ρ. Substituting the observed Fourier norm F_{obs} for $|F|$, we obtain $F'(k) = F_{obs}e^{i\Phi(k)}$, where Φ is the phase of F. Next, ρ' is obtained by the inverse transform of F'. By substituting the observed object norm ρ_{obs} for $|\rho'|$, we have an updated $\rho(r) = \rho_{obs}e^{i\vartheta(r)}$, where ϑ is the phase of ρ'. Let n be an iterative index in the diagram. Then the above procedure can be presented by the iteration $\rho_n \to F_n \to F'_n \to \rho'_n \to \rho_{n+1}$. If the difference $|\rho_n - \rho_{n+1}|$ decreases and becomes to be zero, the object satisfying both constraints is obtained. This iterative procedure is called GS algorithm. The constraints on the usage of the two intensity measurements correspond the electron microscope, and both domain constraints satisfying the experimental settings should be provided. The algorithm corresponds all cases of the reconstruction problem using the measurements in the two domains connected with the conjugate transform.

After a decade of the GS algorithm, Fienup presented the relation between the update of the algorithm and the steepest descent in 1982.[11]

$$\frac{\partial}{\partial\rho} \sum_{k\in K} ||F(k)| - |F'(k)||^2 = 2(\rho - \rho'). \tag{4.1.8}$$

The constraint used in the above equation states that the target is real and positive. The update ρ_n to ρ_{n+1} is referred to as the error reduction (ER) algorithm as follows.

$$\rho_{n+1} = \begin{cases} \rho'_n & x \notin D \\ 0 & x \in D, \end{cases} \tag{4.1.9}$$

where D is the region breaking the object-domain constraint. As another update, the hybrid input-output (HIO) algorithm was introduced in the following.

$$\rho_{n+1} = \begin{cases} \rho'_n & x \notin D \\ \rho_n - \beta\rho'_n & x \in D, \end{cases} \tag{4.1.10}$$

where β is a positive constant. Additionally, an appropriate setting has been used. This algorithm is a type of penalty for the domain D. The fitting degree for the Fourier-domain constraint is presented by the discrimination between the hypotheses $|F_{cal}(k)|$ and $F_{obs}(k)$, and the following R-factor is used.

$$R(F_{cal}, F_{obs}) = \frac{\sum_{k \in K} ||F_{cal}(k)| - |F_{obs}(k)||}{\sum_{k \in K} |F_{cal}(k)|}. \tag{4.1.11}$$

The ER and HIO have been effectively used in diffractive imaging from the experimental data. Miao (1999) has effectively shown the practical usage of these algorithms.[9] Furthermore, the theoretical aspect of the algorithms is still important. Elser provided a differential map that gave the mathematical formulation of the algorithms.[11]

Giving an accurate shape of the target object is difficult in nanoscale or atomic imaging. Tight object support means the region in the object domain which is the positive intensity. Such support is part of important prior information of the Fourier phase retrieval, which is used as an object-domain constraint. Then a method finding the support is required. In the field of crystallography, an effective method for finding the support has been presented by Ozlányi and Säuto. Underlying the object-domain constraint is that the target object is real and positive. The method introduces the sign flip of the object ρ which is applied to the object-domain constraint.

$$\rho_1(x) = \begin{cases} \rho(x) & \text{if } \rho(x) \geq \delta \\ 0 & \text{otherwise,} \end{cases} \tag{4.1.12}$$

$$\rho_2(x) = \begin{cases} \rho(x) & \text{if } \rho(x) < \delta \\ 0 & \text{otherwise,} \end{cases} \tag{4.1.13}$$

Using $\rho_1 - \rho_2$ and $\rho_1 + \rho_2$ based on the threshold δ, the update from ρ_n to ρ_{n+1} is presented in the following update.

$$\rho_{n+1}(x) = \begin{cases} \rho'_n(x) & \text{if } \rho'_n(x) \geq \delta \\ -\rho'_n(x) & \text{if } \rho'_n(x) < \delta, \end{cases} \tag{4.1.14}$$

This method is called "charge flipping", and the threshold is appropriately determined.[12]

4.1.3.3. *Phase retrieval problem in Lagrange form*

The phase retrieval problem is a type of an optimization problem, and a simple estimation of an unknown function using the given constraints. The fundamental formulation of the problem requires the relationship between the prior and posterior objects. These two objects are defined in the object domain and correspond to the prior and obtained objects, respectively, after the calculation of the GS diagram. As the relationship between two terms has been used in information theory, the discrimination is called the information measure, which is the difference between two probability distributions. Using the measure, the algorithm of the phase retrieval problem is presented next.[13]

Let X and Y be the discrete sets corresponding to the object and Fourier domains, respectively, and let ρ and F be the object function and its Fourier transform, respectively. These are the complex-valued functions. Let $\mathcal{P}$ be a set of the functions with a finite volume in the object domain X. The relationship based on the Fourier transform between the object and Fourier domains is established for $\mathcal{P}$. We use the object-domain constraint with real functions and positivity, then define $\bar{\mathcal{P}}$ as a set of all finite-volume and non-negative real functions on X. We introduce the following discrimination between the objects

ρ and τ,

$$D_\gamma(\rho, \tau) = \sum_{x \in X} \left\{ \frac{1}{\gamma} \rho(x)(\rho(x)^\gamma - \tau(x)^\gamma) - \frac{1}{\gamma+1}(\rho(x)^{\gamma+1} - \tau(x)^{\gamma+1}) \right\},$$

(4.1.15)

where $\gamma \in [0, 1]$. This discriminant measure is called density-power divergence, ρ and τ are the probability distribution. Some statistical properties were presented by Basu.[14] The measure is a type of extended f-divergence by Csiszár's work in 1967.[15]

F is the Fourier transform of ρ, and F' is obtained using the observed intensity F_{obs} and the phase of F. The squared error has been used as the discrimination between and F'.

$$E(F, F') = \sum_{k \in K} |F(k) - F'(k)|^2.$$

(4.1.16)

The phase of F' is same as that of F. Thus $E(F, F') = E(|F|, |F'|)$. Using the discrimination D_γ, we obtain the following equation.

$$\lim_{\gamma \to 1} D_\gamma(F, F') = E(F, F').$$

(4.1.17)

The difference between ρ and ρ' is required as well as between F and F'. Introducing the discriminant measure D_γ, the GS-diagram is extended as the Fourier and its inverse transforms between two function spaces shown in Fig. 4.1.8.

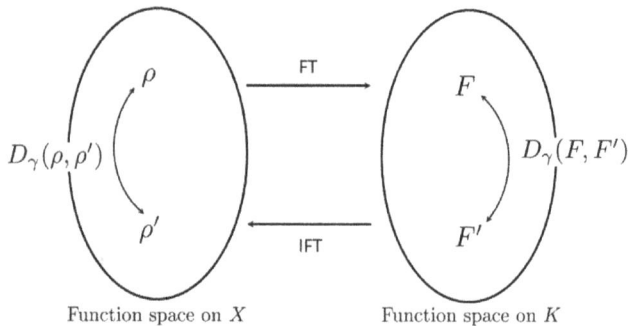

Fig. 4.1.8 Information measures and GS-diagram.

The mathematical formulation of the phase retrieval algorithms requires how to obtain the posterior τ when prior ρ is given. When the discriminant measures of two spaces are sufficiently small, the posterior is a plausible phase-retrieved object. Then, the sum of two discriminant measures is introduced as the Lagrange form.

$$L(\lambda) = D_\gamma(\rho, \tau) + \lambda E(F, F'). \tag{4.1.18}$$

By $\frac{\partial L(\lambda)}{\partial \rho} = 0$, we obtained the following equation.

$$\frac{\rho(x)^\gamma - 1}{\gamma} = \frac{\tau(x)^\gamma - 1}{\gamma} + \lambda(\rho'(x) - \rho'(x)). \tag{4.1.19}$$

Assuming that $|\rho - \tau|$ is sufficiently small and with other approximations, the update from ρ_n to ρ_{n+1} is expressed as follows.

$$\rho_{n+1}(x) = \left\{\rho_n(x)^\gamma + C_\gamma(\rho'_n(x) - \rho_n(x))\right\}^{\frac{1}{\gamma}}. \tag{4.1.20}$$

In the case of $\gamma = 1$, then we have

$$\rho_{n+1}(x) = (1 - C_\gamma)\rho_n(x) + C_\gamma \rho'_n(x), \tag{4.1.21}$$

and $C_\gamma = 1$, this is equivalence to the ER algorithm. When γ is close to zero, we have

$$\rho_{n+1}(x) = \rho_{n+1}(x)e^{C_\gamma(\rho'_n(x) - \rho_n(x))} \tag{4.1.22}$$

In this case, the discrimination measure is presented in the following.

$$\lim_{\gamma \to 0} D_\gamma(\rho, \tau) = \sum_{x \in X} \rho(x) \ln \frac{\rho(x)}{\tau(x)} + \sum_{x \in X} \tau(x)$$
$$- \sum_{x \in X} \rho(x) \tag{4.1.23}$$

This measure is called I-divergence,[16] and when ρ and τ are the probability distributions, the measure is called Kullback-Leiber divergence,[17] which is well used in information theory.

The maximum entropy method (MEM) is widely used in science and engineering fields. MEM crystallography was introduced by Collins,[18, 19]

r and k are the indexes of the object and Fourier domains, respectively. The entropy function by Jaynes[20] was introduced as follows.

$$S(\bar{\rho}, \bar{\tau}) = -\sum_r \bar{\rho}(r) \ln \frac{\bar{\rho}(r)}{\bar{\tau}(r)}, \tag{4.1.24}$$

$$\bar{\rho}(r) = \frac{\rho(r)}{\sum_{r'} \rho(r')}, \quad \bar{\tau}(r) = \frac{\tau(r)}{\sum_{r'} \tau(r')}, \tag{4.1.25}$$

$\rho(r)$ and $\tau(r)$ are the estimated and prior distributions, respectively, which correspond to the electron density function. $\bar{\rho}(r)$ and $\bar{\tau}(r)$ are normalized by their volume. The Fourier-domain constraint comprises the known and unknown phase is given as

$$C_1 = \frac{1}{M_1} \sum_k \frac{|F_{cal}(k) - F_{obs(k)}|^2}{\sigma(k)^2}, \tag{4.1.26}$$

$$C_2 = \frac{1}{M_2} \sum_k \frac{||F_{cal}(k)| - |F_{obs}(k)||^2}{\sigma(k)^2}, \tag{4.1.27}$$

where M_1 and M_2 are the reflection number, and F_{cal} is the calculated structure factor as in the following.

$$F_{cal}(k) = V \sum_r \rho(r) e^{-i2\pi rk}, \tag{4.1.28}$$

where V is the volume of unit cell, F_{obs} is the observed structure factor, and $\sigma(k)^2$ is the variance of F_{obs}. Based on the constraint of these structure factors, the following is introduced for maximizing the entropy.

$$L_\lambda = S(\rho, \tau) - \frac{\lambda_1}{2} C_1 - \frac{\lambda_2}{2} C_2. \tag{4.1.29}$$

The maximum entropy distribution is obtained by the following.

$$\rho(r) = \exp\left\{ \ln \tau(r) + \frac{\lambda_1 F_0}{M_1} \sum_k \frac{(F_{obs}(k) - F_{cal}(k))}{\sigma(k)^2} \exp\{-i2\pi kr\} \right.$$

$$\left. + \frac{\lambda_2 F_0}{M_2} \sum_k \frac{(|F_{obs}(k)| \exp[i\psi(k)] - F_{cal}(k))}{\sigma(k)^2} \exp\{-i2\pi kr\} \right\}, \tag{4.1.30}$$

where F_0 is the total number of electrons in a unit cell, and ψ is a model phase. The MEM electron density function is obtained using F_{obs}, σ, λ_1 and λ_2. The difference and correspondence between crystallography and phase retrieval were investigated by Millane in 1990.[21]

A relationship between the MEM crystallography and the phase retrieval based on information measures are described in the following. We introduce the measure, I-divergence Eq. (4.1.23). In the phase retrieval problem, the Fourier phase is unknown, and then the known structure factor is ignored in the MEM formula, and the phase of F_{cal} is used for F_{obs}. ρ' is to be the inverse Fourier transform of $|F_{\text{obs}}|\exp[i\psi]$, where ψ is settled by the phase of F_{cal} in here. Then we have

$$\rho(r) = \exp\{\ln \tau(r) + C_{\text{mem}}(\rho'(r) - \rho(r))\}, \qquad (4.1.31)$$

where C_{mem} is the positive constant obtained by the reduction of the maximum entropy distribution of crystallography. This result is the same as the case of the discriminant measure D_γ with $\gamma \rightarrow 0$, and then MEM crystallography is related to phase retrieval based on the information measure analysis.[22]

Fienup's work has been used as a method for reconstructing the phase from intensity measurements. Even if the mathematical analysis is developed in phase retrieval, such the simple usage will continue in the future.

References

[1] N. Tanaka, *Electron Nano-imaging*, Springer, Tokyo, 2017.
[2] J. Yamasaki, *et al.*, *Appl. Phys. Lett.* **101**, 234105 (2012).
[3] S. Morishita, *et al.*, *Appl. Phys. Lett.* **93**, 183103 (2008).
[4] S. Morishita, *et al.*, *AMTC Letters* **2**, 116–117 (2010).
[5] O. Kamimura *et al.*, *Ultramicroscopy* **110**, 130 (2010).
[6] O. Kamimura *et al.*, *Appl. Phys. Lett.* **98**, 174103 (2011).
[7] K. Kawahara *et al.*, *Phys. Rev. B*. **81**, 081404 (2010).
[8] R. W. Gerchberg and W. O. Saxton, *Optic*, **35**, 237–246 (1972).
[9] J. Miao, *et al.*, *Nature*, **400**, 342–344 (1999).
[10] J. R. Fienup, *Appl. Optics* **21**, 2758 (1982).
[11] V. Elser, *J. Opt. Soc. Am. A* **20**, 40 (2003).
[12] G. Oszlányi and A. Säuto, *Acta Cryst.* **A60**, 134 (2003).
[13] H. Shioya and K. Gohara, *Optics Commun.* **266**, 88 (2006).
[14] A. Basu, *et al.*, *Biometrika* **85**, 549 (1998).
[15] I. Csiszár, *Studia Sci. Math. Hunger.* **2**, 299–318 (1967).

[16] I. Csiszár, *Ann. Math. Stat.* **19**, 2033 (1991).

[17] S. Kullback and R. A. Leibler, *Ann. Math. Stat.* **22**, 79 (1951).

[18] D. M. Collins, *Nature* **298**, 49 (1982).

[19] M. Sakata and M. Sato, *Acta Cryst.* **A46**, 263 (1990).

[20] E. T. Jaynes, *IEEE Trans SystCybern* **SSC4**, 227 (1968).

[21] R. P. Millane, *J. Opt. Soc. Am. A* **7**, 394 (1990).

[22] H. Shioya and K Gohara, *J. Opt. Soc. Am. A* **25**, 2846 (2008).

NANOSTRUCTURES AND SINGLE MOLECULE IMAGING

Hiroshi Sekiguchi

Japan Synchrotron Radiation Research Institute

Yuji C. Sasaki

The University of Tokyo

4.2.1. Single Molecule Imaging Using Quantum Beams

4.2.1.1. *Necessity of single molecule imaging*

In the research areas dealing with organic compounds and bio materials, it would be one of goal for researchers to determine a stable 3D-structure of the target molecule, and to capture its internal motion at single molecule level because such information is important to clarify the mechanism of its function. Single molecule measurement technology was emerged from the time when T. Hirschfeld showed that the number of fluorescent molecule on the substrate surface was countable in 1976.[1] The technology is advanced and matured to have research area, such as "single molecule measurements", "single molecule physiology" and etc., and is applied to surface science and polymer science area. Much of efforts were put for improving sensitivity and accuracy to detect fluorescence from target molecule in visible light microscopy. Since proteins are major component in biological systems, it would be one of best strategy

visible light (λ= 300-800 nm)

Imaging concept
　diffraction limits (λ/2)

Tracking concept
　　　λ/100

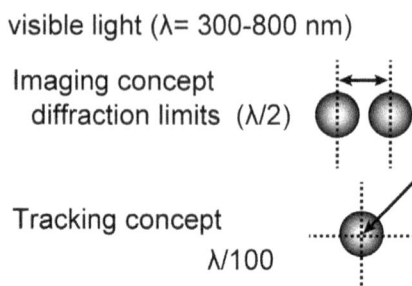

Fig. 4.2.1　Schematics of Imaging and tracking concept.

for understanding mechanism of bio-system to establish a methodology which enable to get target protein's internal dynamic information at single molecule level with micro seconds time resolution. And it would be noted that such method could be applied *in vivo* and to variety of proteins.

Single molecule technology and the concept of a microscope are fundamentally different. "Single molecule imaging" that deviates from the common sense of the optical referred to as "diffraction limit" was an epoch-making idea. The idea is basis of Super-Resolution technology, which was awarded the Nobel Prize in Chemistry in 2014.[2] Figure 4.2.1 is a conceptual diagram of single molecule technology. The technology beyond the "diffraction limit" was achieved by not only hardware improvements but also by latest image analysis techniques. Thus, in the future development of measurement technology, it is required to develop both hardware and software which adapt artificial intelligence.

4.2.1.2.　*Advantages and disadvantages of using labeling method*

In the quantum beam utilization research area, the idea of "tracer" than the term "label" have been used. This idea has long history; the first proposal is a radioactive tracer method. The biological research using 32P by G. Hevesy and O. Chiewitz who are well-known for using artificial radioisotope.[3] H. Blumgart, is called the father of nuclear medicine, utilized radium to measure the condition of patients with heart failure.[4] His statements at the time for tracer's features was as follows, (1) it does not present in the body, (2) it does not affect the behavior to be inspected, (3) it must be a substance that can be detected even in very small amounts, and (4) it would be disappeared from the body at an appropriate speed. Currently, the number of available use of radioactive tracers from the characteristics of the radioactive is depleted every year. Now, sensitive fluorescent molecules or quantum dots are used for tracers instead of radioisotopes. The green fluorescent protein (GFP) discovered by Prof. O. Shimomura is one of most efficient tracer molecule *in vivo*[5] and he received Nobel Prize in Chemistry in 2008. The other fluorescent molecules have four characteristics of the tracer as mentioned above and currently are to be commercially available as many radioisotopes were presented in 1920–1940s.

4.2.1.3. *Diffracted X-ray tracking method*

The mechanism of protein's function would be understood if the dynamic information at atomic level with real time, micro seconds level. Such subtle information would be available with probes in same order of wavelength, X-ray. However, many researchers are wondering whether single molecule measurements are achieved with X-ray. It is the common sense that the magnitude of scattering cross-section is proportional to the wavelength. As most of single molecule technology utilizes visible light, X-ray would be difficult as probe even brilliant X-ray light source are available with synchrotron facility. Diffracted X-ray tracking (DXT) is a new measurement technique that utilizes most high sensitivity phenomenon of X-ray diffraction. The concept of the method is simple (Fig. 4.2.2).[6,7] The nanocrystal, motion tracer, having a diameter of about 20 nm was labeled on protein molecule, and the

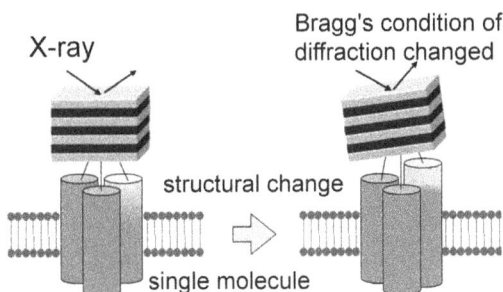

Fig. 4.2.2 Schematics of diffracted X-ray tracking method.

trajectory of diffraction spots from the nanocrystal was investigated as motion of the protein with time-resolved, micro — or milli seconds scale, diffraction images. Figure 4.2.2 shows that schematic diagrams in DXT for structural changes of protein molecule synchronizes the rotational motion of nanocrystal labelled on the protein that reflects diffraction spots motion in time-resolved diffraction images. Please note that the information detected in DXT is the rotational motion of nanocrystal on the protein, and not the translational motion of the nanocrystal. And DXT analysis is on the assumption for nanocrystal on the protein that is not interfere the original motion of the protein and that is transmitted the feature of the motion of the target protein. The labelling condition of the nanocrystal on the protein depends the success rate of DXT measurement, many try-and-error DXT measurement's result showed that DXT works in a relatively wide range of experimental conditions.

As described above, the diffracted X-ray tracking method is not the X-ray single molecule detection methods, but is the X-ray labeling technique for internal motion analysis of the target protein molecule. This technique is different from the single molecule analysis with X-ray free electron laser (XFEL). Single particle analysis using electron beam[8] and magnetic resonance force microscope combining with computational science technique for structure determination are extremely inevitable, and X-ray field was behind in such technology.

Figure 4.2.3 shows why it is possible to achieve finer accuracy than the atomic size. The concept of DXT is already shown in Fig. 4.2.1 that describes "molecular tracking concept" that is beyond "diffraction limits". In 1998, it was invented that diffracted X-ray tracking method that is extended super resolution concept in X-ray region.[6] Since two rotational dynamic information, θ and χ directions, could be investigated independently in DXT, 3-dimensional dynamic information is reconstructed considering the polar coordinates (Fig. 4.2.4). If the protein molecule is inserted perpendicularly to the substrate and lipid bilayer (membrane protein's case), tilting motion reflects the diffraction spot movement in θ directional motion in diffraction images and rotational motion on substrate of membrane surface is reflect to the χ directional motion in diffraction images. Much of quantitative analysis for proteins in DXT are

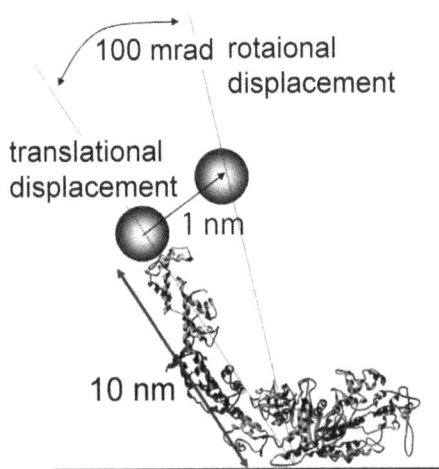

Fig. 4.2.3 If the molecule with 10 nm in length rotates 0.1 mrad on the edge of the molecule, the length of translational motion of the molecule would be 1 pm. Such subtle motion could be measured by X-ray diffraction method.

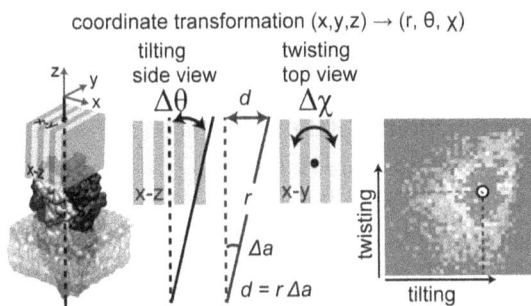

Fig. 4.2.4 3D internal motion could be analyzed from angular displacement in both the tilting and twisting directions in fixed time interval. Such motion is summarized by 2D angular displacement histograms.

needed to probe that nanocrystal which is labelled on the protein could follows the functional motion of proteins.[9–11]

4.2.1.4. *Single molecule imaging using next generation quantum beam*

DXT can be said to be a time-resolved Laue diffraction method from a tracer. Diffraction using quantum beam from the tracer is not a

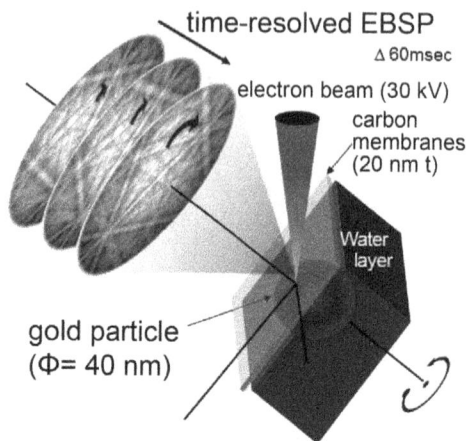

Fig. 4.2.5 Schematics of diffracted electron tracking time-resolved EBSP (electron backscatter pattern).

phenomenon unique to X-rays. The concept of DXT could be applied another quantum beams, such as electron beam or neutron beam. Diffracted electron tracking (DET) is already achieved using electron backscattering diffraction (EBSD) with electron beam.[12] Figure 4.2.5 shows that EBSD pattern and DET setups. Please note that the damage of electron beam for the sample is extremely severe, therefore the beam should be small enough not be irradiating the sample. It is one of attractive side of DET that DET measurement could be realized in a laboratory scale, meanwhile DXT needs synchrotron radiation facility. Because of X-ray's transparency for the sample, DXT could target not only to the sample surface but to the inner side of the sample *in vivo* condition. We should select proper quantum beams considering the objective of the measurement, such as functional analysis of the target protein *in vivo* condition.

For the outlook for future measurement of biological sample, diffracted neutron tracking (DNT) will be important because of neutron beams' nondestructive aspect. The long-term measurement, several days' scale, would be important for tissues and individuals, that are assembles from several cells, therefore non-destructive probe of neutron beam would be essential such measurements.

Advanced single molecule measurement technique, such as combination of DXT with the patch clamp technique and DXT with visible-light single molecule techniques, will be applied to the structure kinetics measurement of intracellular channel protein molecule, dynamic measurement of intrinsically disordered protein molecules, local viscosity measurements for supersaturated solution, and dynamic fluid mechanics in microfluidic devices. Additional X-ray interference phenomenon or evanescent wave of X-ray total reflection phenomenon would be attractive and such phenomenon would be applied for the principles of single molecule detection if the phenomenon is enough sensitive to realize high speed measurements.

References

[1] T. Hirschfeld, *Appl. Opt.* **15**, 2965 (1976).
[2] C. Eggeling, *et al.*, *Nature* **457**, 1159 (2009).
[3] O. Chiewitz and G. Hevesy, *Nature* **136**, 754 (1935).
[4] H. Blumgart and O. C. Yens, *J. Clin. Invest.* **4**, 1 (1926).
[5] O. Shimomura, F. H. Johnson, and Y. Saiga, *J. Cell. Comp. Physiol.* **59**, 223 (1962).
[6] Y. C. Sasaki, *et al.*, *Phys. Rev. E.* **62**, 3843 (2000).
[7] Y. C. Sasaki, *et al.*, *Phys. Rev. Lett.* **87**, 248102 (2001).
[8] T. Nakagawa, *et al.*, *Nature* **433**, 545 (2005).
[9] H. Shimizu, *et al.*, *Cell* **132**, 67 (2008).
[10] Y. C. Sasaki, Dynamical Observations of Soft Nanomaterials Using X-rays or High-energy Probes, p. 69–107, SOFT NAMOMATERIALS, American Scientific Publishers (2009).
[11] Y. C. Sasaki, Picometer-scale Dynamic X-ray Imaging, 209–234, Fundamentals of Picoscience, CRC Press (2013).
[12] N. Ogawa, *et al.*, *Sci. Rep.* **3**, 2201 (2013).

Chapter 5

FIRST-PRINCIPLES ELECTRONIC STRUCTURE CALCULATION AND MOLECULAR DYNAMICS SIMULATION

Yoshitada Morikawa

Osaka University

Yu Takano

Hiroshima City University

It is becoming possible by the rapid development of computer simulation techniques in recent years to predict the atomic structures, electronic and magnetic properties, mechanical strength, chemical reactivity, etc. of materials from theoretical simulations. It is also possible to elucidate the physical factors that govern them, and based on this knowledge, it has also been tried to give guidelines for designing new materials with more desirable properties using computer simulations without conducting experiments.

In order to predict the structures of materials, it is necessary to calculate the interaction energy between atoms and the stability (energy or free energy) with different atomic arrangement accurately. There are various methods for calculating the interaction energy between atoms, from high to low accuracy. There is a trade-off relation between the calculation accuracy and the scale (number of atoms) of the calculation model that can be handled, namely the number of atoms that can be handled is small in the method with high calculation accuracy, whereas the method with reduced calculation accuracy can calculate large scale systems. In this chapter, we will explain a first-principles electronic structure calculation method which is representative as a method with high calculation precision and the molecular dynamics simulation which can calculate a large scale at high speed although calculation accuracy is considerably lower.

5.1.   Electronic Structure Calculation Methods

5.1.1.   *Review of quantum mechanics*

In order to understand the behavior of electrons in materials, it is required to calculate the electronic state using the Schrödinger equation, which is the fundamental equation of quantum mechanics.[1] At first, let us review the quantum mechanics briefly. A wave function $\psi_i(\mathbf{r})$ of an electron with the mass m in the potential $V(\mathbf{r})$ satisfies the following Schrödinger equation.

$$
\begin{aligned}
\hat{\mathcal{H}}\psi_i(\mathbf{r}) &= -\frac{\hbar^2}{2m}\Delta\psi_i(\mathbf{r}) + V(\mathbf{r})\psi_i(\mathbf{r}) \\
&= -\frac{\hbar^2}{2m}\left(\frac{\partial^2\psi_i(\mathbf{r})}{\partial x^2} + \frac{\partial^2\psi_i(\mathbf{r})}{\partial y^2} + \frac{\partial^2\psi_i(\mathbf{r})}{\partial z^2}\right) + V(\mathbf{r})\psi_i(\mathbf{r}) \\
&= E_i\psi_i(\mathbf{r}),
\end{aligned}
\tag{5.1}
$$

where $\mathbf{r} = (x, y, z)$ is a position vector. The energy and the electron density distribution, $\rho_i = |\psi_i(\mathbf{r})|^2 = \psi_i^*(\mathbf{r}) \cdot \psi_i(\mathbf{r})$, of this system can be estimated by using the i-th energy eigenvalue, E_i and the corresponding wave function, $\psi_i(\mathbf{r})$, which are obtained by solving this equation.

A hydrogen atom is the basis of the electronic state of materials, and it is composed of a proton having a charge of $+e$ and an electron having a charge of $-e$. Substitution of the Coulomb interaction between a proton and an electron, $V(\mathbf{r}) = -e^2/4\pi\varepsilon_0 r$, for Eq. (5.1) yields the Schrödinger equation of a hydrogen atom. We can analytically solve this equation and obtain the energy eigenvalues and the corresponding wave functions of an electron in a hydrogen atom. The wave functions can be expressed by a product of a radial wave functions, $R_{nl}(r)$, and spherical harmonics, $Y_l^m(\theta, \varphi)$, as $\psi_i(\mathbf{r}) = R_{nl}(r)Y_l^m(\theta, \varphi)$, where n, l, and m are called principal quantum numbers with $n = 1, 2, 3, \ldots$, azimuthal quantum numbers with $l = 0, 1, \ldots, n-1$, and magnetic quantum numbers with $m = -l, -l+1, -l+2, \ldots, l-1, l$, respectively. The principal quantum number determines the energy eigenvalues of an electron in a hydrogen atom: $E_{nlm} \sim -1/n^2$. In addition, it specifies the spatial distribution of electrons with respect to the distance from the nucleus. In contrast, the azimuthal quantum number specifies the azimuthal distribution of electrons with respect to the angles, θ and φ. The wave function with azimuthal quantum numbers (l) of 0, 1, 2, 3 are called s-orbital, p-orbital, d-orbital, and f-orbital, respectively.

5.1.2. *Ab initio molecular orbital calculation*

In recent years, computational methods called "first-principles electronic structure calculation" or "first-principles calculation" have attracted attention in the field of material science. "First-principles calculation" means a method calculating the electronic structure based on quantum mechanics without using experimental data other than atomic numbers, and has been developed in the field of solid state physics. Even in the field of molecular science, "*ab initio* molecular orbital calculation"[2–4] has been developed for electronic structure calculations of a molecule without using empirical parameters that reproduce the experimental results. Both "first-principles calculation" and "*ab initio* molecular orbital calculation" aim to accurately calculate the electronic state of materials and to theoretically predict their properties. They are closely related to each other but have been developed independently, because of different targets of computations (solids in the first-principles calculations, molecules in the *ab initio* molecular orbital calculations).

Let us consider the electronic state of a molecule which consists of N_{atom} nuclei and $N_{electron}$ electrons. In electronic structure calculations of molecules and solids, we can calculate the electronic states in stationary nuclei since nuclei are much heavier than electrons. This is called "adiabatic approximation" or "Born–Oppenheimer approximation". Schrödinger equation for electrons in this system is expressed by the following equation.

$$\hat{\mathcal{H}}\Psi(\mathbf{x}_1, \mathbf{x}_2, \ldots, \mathbf{x}_{N_{electron}}) = E\Psi(\mathbf{x}_1, \mathbf{x}_2, \ldots, \mathbf{x}_{N_{electron}}), \tag{5.2}$$

where $\mathbf{x}_i$ is the coordinate of the i-th electron indicating the combination of a spatial coordinate vector, $\mathbf{r}_i$, and a spin coordinate, ξ_i. $\hat{\mathcal{H}}$ is the Hamiltonian of the system and expressed as follows.

$$\hat{\mathcal{H}} \equiv \sum_{i=1}^{N_{electron}} -\frac{\hbar^2}{2m}\Delta_i - \sum_{i=1}^{N_{electron}} \sum_{A=1}^{N_{atom}} \frac{Z_A e^2}{4\pi\varepsilon_0 |\mathbf{r}_i - \mathbf{R}_A|}$$

$$+ \sum_{i=1}^{N_{electron}} \sum_{j>i}^{N_{electron}} \frac{e^2}{4\pi\varepsilon_0 |\mathbf{r}_i - \mathbf{r}_j|} \tag{5.3}$$

In Eq. (5.3), $\mathbf{R}_A$ is the position vector of the nucleus A, and $Z_A e$ denotes the charge of the nucleus A. The first term of Eq. (5.3) is the operator for the kinetic energies of the electrons; the second term represents the Coulomb interactions between electrons and nuclei; the third term represents the Coulomb interactions between electrons. By solving this, the electronic energy of the whole system, $E(\mathbf{R}_1, \mathbf{R}_2, \mathbf{R}_3, \ldots \mathbf{R}_{N_{atom}})$ is obtained as a function of the position of atoms. Using the energy depending on the position of the atoms, we can find a stable molecular geometry, a chemical reaction pathway with its energy barrier, and an electron-transfer pathway in the electron transfer reaction. Besides, it is also possible to determine the parameters required for molecular dynamics (MD) simulations.

When computing the electronic structure of a molecule, we usually use atomic orbitals as basis sets rather than plane waves used in band calculations of solids. Molecular orbitals are constructed from atomic orbitals, and the Schrödinger equation is approximately solved by using it. This is referred to as the molecular orbital method. Atomic orbitals

(a) **Atomic orbital**

s orbital p orbital   d orbital

(b) **Molecular orbital**

linear combination of p orbitals

$$= C_1 \times \text{[}} + C_2 \times \text{[}} + C_3 \times \text{[}} + C_4 \times \text{[}}$$

HOMO of 1,3-butadiene

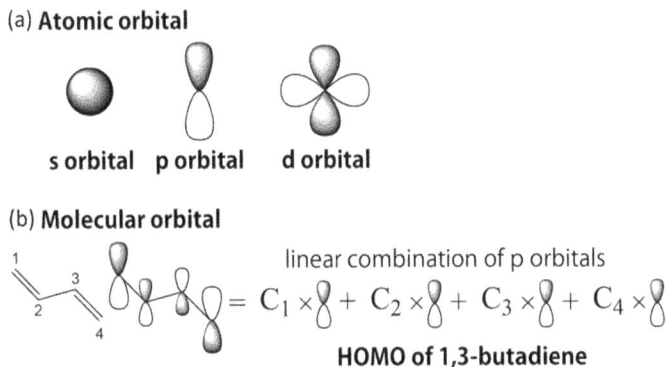

Fig. 5.1   Atomic orbitals and molecular orbitals.

are one-electron wave functions obtained by solving the Schrödinger equations for an atom. The atomic orbitals are named as s-orbitals ($l = 0$), p-orbitals ($l = 1$), and d-orbitals according to the azimuthal quantum number, l (Fig. 5.1(a)). Molecular orbital represents a one-electron state in the molecule and is formed by linear combination of atomic orbitals (LCAO). Figure 5.1(b) shows the highest occupied molecular orbital (HOMO) of 1,3-butadiene as an example of a molecular orbital. You can see that the HOMO is made from the atomic orbitals (p-orbitals) of carbon atoms constituting 1,3-butadiene. Although it is necessary to solve the Schrödinger equation for molecules to obtain molecular orbitals, we cannot solve it exactly due to the electron–electron interactions. Hence, instead of explicit treatment of electron-electron interactions, we calculate the electronic state in the potential which is an average of the potentials created by all the other electrons, as an approximation (independent particle approximation or mean field approximation). This is called the Hartree–Fock (HF) method. However, electrons repel each other according to the Coulomb repulsion. This is called electronic correlation, and it must be considered for the reliable and accurate electronic state calculation. Therefore, in order to correct the wavefunctions obtained by the HF method, the Configuration Interaction method, the Møller–Presset many-body perturbation method, and the Coupled Cluster method have been developed. However, the computational costs for these methods are so expensive that applications to large molecules are limited.

Another method for treating electronic correlation is the density functional theory (DFT) mentioned in the later section. In this method, the Kohn–Sham equation equivalent to the Schrödinger equation is solved under the Hohenberg–Korn theorems that electron density determines the properties of the molecule. The advantage of the DFT is that the computational cost is comparable to that of the HF method even though it somewhat incorporates the electronic correlation. DFT is now one of the most popular computational procedures for molecular electronic structure calculations. Electronic correlation in DFT is involved in the exchange–correlation functional. Since the exact form is not known, a number of approximate exchange–correlation functionals have been developed. It is known "B3LYP" as the most popular exchange-correlation functional to reproduce experimental values of molecular properties. However, since the electronic structure of a metal active site in a metalloprotein is complicated, it is indispensable to verify the validity of the exchange–correlation functional in the electronic structure calculations of the metal active site.

Table 5.1 shows popular software packages of molecular orbital calculations. In particular, Gaussian and GAMESS are most often used in molecular orbital calculations. Gaussian implements various functions necessary for the electronic structure calculations of a molecule, and has a user-friendly interface, GaussView. A wide range of quantum chemical computations are possible by using GAMESS. A site license for GAMESS is available at no cost to both academic and industrial users. Other research groups develop user-friendly interface such as wxMacMolPlt (https://brettbode.github.io/wxmacmolplt/) and Facio (http://zzzfelis.

Table 5.1 Popular software packages for molecular orbital calculations and their website.

Software packages	Website
Gaussian	http://gaussian.com/
GAMESS	http://www.msg.ameslab.gov/gamess/
NWChem	http://www.nwchem-sw.org/index.php/Main_Page

sakura.ne.jp/). NWChem provides many methods to compute the properties of molecular and periodic systems and is designed for efficient execution on massively parallel computers. The code is distributed as open-source under the terms of the Educational Community License version 2.0.

5.1.3. *Density functional theory*

While non-empirical molecular orbital methods have been developed in the field of molecular science, electronic state calculation methods based on density functional theory have been developed in the field of solid state physics. In *ab initio* molecular orbital methods, wave functions of molecules are determined accurately to calculate various properties of molecules. Therefore, they are also called "wave function theory". On the other hand, in the density functional theory, it is a method to calculate various properties of solids by accurately obtaining "electron density".[5] Based on the Hohenberg-Korn theorem, it is proved that various physical quantities, including the total energy of the ground state, can be uniquely determined from the electron density instead of wave functions. In usual quantum mechanics, given the external field $V(\mathbf{r})$ (for example, the electric field created by a nucleus with positive charge), we solve the Schrödinger equation (5.3) and find the ground state wave function $\Psi(x_1, x_2, \ldots, x_{N_{\text{electron}}})$. Once the wave function is found, it is possible to calculate the expectation values of various physical quantities. For example, in the case of total energy, it can be calculated as the expectation value of the Hamiltonian.

$$E = \int dx_1 \int dx_2 \cdots \int dx_{N_{\text{electron}}} \Psi^*(x_1, x_2, \ldots, x_{N_{\text{electron}}})$$

$$\times \hat{\mathcal{H}}\Psi(x_1, x_2, \ldots, x_{N_{\text{electron}}}) \tag{5.4}$$

The electron density can also be obtained from the following formula.

$$\rho(\mathbf{r}) = \int d\xi \int dx_2 \cdots \int dx_{N_{\text{electron}}} |\Psi(x, x_2, \ldots, x_{N_{\text{electron}}})|^2 \tag{5.5}$$

On the other hand, according to Hohenberg-Korn theorem, given the electron density $\rho(\mathbf{r})$, it was shown that the external field $V(\mathbf{r})$ giving the

Wave function theory

$$V(\mathbf{r}) \longrightarrow \Psi(\mathbf{r}) \longrightarrow \rho(\mathbf{r}), \quad \langle \Psi | \hat{A} | \Psi \rangle$$

External potential   Wave function   Density, physical observables

Density functional theory

$$\rho(\mathbf{r}) \longrightarrow V(\mathbf{r}) \longrightarrow \Psi(\mathbf{r}), \quad \langle \Psi | \hat{A} | \Psi \rangle$$

Density   External potential   Wave function, physical observables

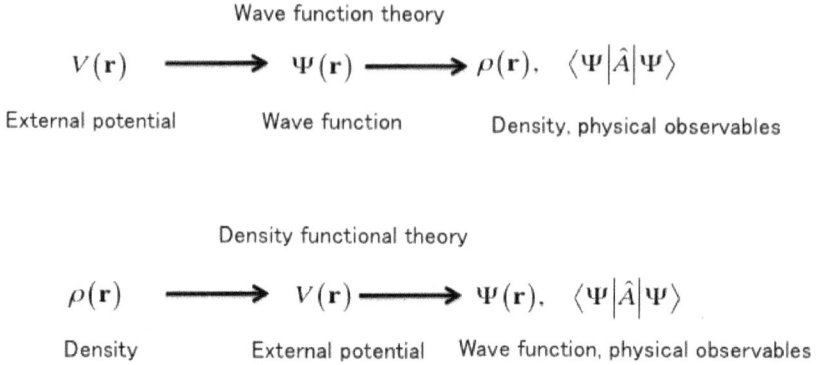

Fig. 5.2   Wave function theory and density functional theory.

electron density as the electron density of the ground state is uniquely determined (Fig. 5.2). Since the wave function depends on the spatial coordinates and the spin coordinates of each electron, it is a complex function of $4N_{electron}$ dimension and it is a very complicated function. Therefore, as the number of electrons increases, the effort to obtain a highly accurate wave function explosively increases. Since the atoms are periodically and infinitely arranged in an ideal solid, the number of electrons is also infinite and it is impossible even to write the wave function explicity, but if the electron density $\rho(\mathbf{r})$, even for a system with infinite number of electrons, it can be written as a periodic function in three-dimensional space, and it can be much simplified. From the electron density, the total energy of the system is calculated as follows

$$E = -\frac{\hbar^2}{2m} \sum_{i=1}^{N_{ectron}} \int \psi_i^*(\mathbf{r}) \Delta \psi_i(\mathbf{r}) - \sum_{A=1}^{N_{atom}} \int \frac{Z_A e^2 \rho(\mathbf{r})}{4\pi\varepsilon_0 |\mathbf{r} - \mathbf{R}_A|} d\mathbf{r}$$

$$+ \frac{1}{2} \iint \frac{\rho(\mathbf{r})\rho(\mathbf{r}')}{4\pi\varepsilon_0 |\mathbf{r} - \mathbf{r}'|} d\mathbf{r} d\mathbf{r}' + E_{xc}[\rho(\mathbf{r})] \tag{5.6}$$

$$\rho(\mathbf{r}) = \sum_{i=1}^{N_{electron}} |\psi_i(\mathbf{r})| \tag{5.7}$$

Here, $\psi_i(\mathbf{r})$ is an auxiliary one-electron wave function introduced to accurately calculate the kinetic energy of electrons. The first term of the Eq. (5.6) is the kinetic energy, the second term is the interaction energy with the nucleus, the third term is the classical Coulomb interaction between electrons (Hartree energy), the fourth term is exchange correlation energy. It is called an exchage-correlation energy functional and represents quantum mechanical correction for Hartree energy. One-electron wave function $\psi_i(\mathbf{r})$ is obtained by solving the following Kohn-Sham equation.

$$\left\{-\frac{\hbar^2}{2m}\Delta - \sum_{A=1}^{N_{\text{atom}}} \frac{Z_A e^2}{4\pi\varepsilon_0|\mathbf{r}-\mathbf{R}_A|} + \int \frac{\rho(\mathbf{r})\rho(\mathbf{r}')}{4\pi\varepsilon_0|\mathbf{r}-\mathbf{r}'|}d\mathbf{r}' + V_{xc}[\rho(\mathbf{r})]\right\}\psi_i(\mathbf{r})$$
$$= \epsilon_i\psi_i(\mathbf{r}) \tag{5.8}$$

Here, V_{xc} is called an exchange correlation potential,

$$V_{xc}[\rho](\mathbf{r}) = \frac{\delta E_{xc}}{\delta\rho(\mathbf{r})} \tag{5.9}$$

Since the electron density $\rho(\mathbf{r})$ included in the Kohn-Sham Eq. (5.8) is calculated from the one-electron wave functions by Eq. (5.7), when solving Eq. (5.8), the trial density $\tilde{\rho}(\mathbf{r})$ is used as an approximate density. After obtaining one electron wave functions, a new density is calculated from Eq. (5.7) from the obtained one-electron wave functions. Solving equation (5.8) using the newly obtained density ... and so on, it is necessary to calculate self-consistently by repeating this procedure.

Although the total energy is given by Eq. (5.6) in the density functional theory, this accuracy is largely dependent on the exchange correlation energy functional $E_{xc}[\rho]$ in the fourth term. Although the exact exchange correlation energy functional should exist, its explicit form is not known. Therefore, it is necessary to calculate approximately. The simplest approximation is an approximation called the Local Density Approximation (LDA), which divides the entire space into small regions, assuming that the electron density is uniform in each region and replacing the energy with that of uniform electron gas of the average density in that region. It approximates the correlation energy with that of the uniform electron gas.

$$E_{xc}^{LDA}[\rho] = \int \epsilon_{xc}^{LDA}(\rho(\mathbf{r}))\rho(\mathbf{r})d\mathbf{r} \tag{5.10}$$

Here, $\epsilon_{xc}^{LDA}(\rho(\mathbf{r}))$ is the exchange correlation energy density of uniform electron gas, and an approximate expression with high accuracy is known as a function of density. LDA is an extremely simple approximation, and the electron density changes abruptly especially in the vicinity of the nucleus, so it seems that the approximation is not good. Actually, however, it was turned out that the band structures and stability of various solids can be calculated accurately using LDA, and therefore, it has been widely used in the calculation of the electronic state of solids. However, several shortcomings were observed, namely, (1) overestimate the cohesive energy of the solid and the interatomic bond energy of the molecule, (2) the dispersion force (long-range van der Waals interaction) is not expressed, (3) energy band gaps of semiconductors and insulators are underestimated. It was shown that the accuracy is significantly improved by Generalized Gradient Approximation (GGA) which incorporates up to the first-order of the electron density gradient in addition to the local density of electrons.

$$E_{xc}^{GGA}[\rho] = \int \epsilon_{xc}^{GGA}(\rho(\mathbf{r}), |\nabla\rho(\mathbf{r})|)\rho(\mathbf{r})d\mathbf{r} \qquad (5.11)$$

It was also proposed to combine the exchange correlation energy by GGA and the correct exchange energy (Hartree-Fock exchange energy) (hybrid functional) and the accuracy has been further enhanced. In comparison with Hartree-Fock + second-order perturbation correction (MP 2), which was used in the field of quantum chemistry, it has been shown that the amount of calculation is much smaller and the accuracy is higher. The hybrid functional also gives improved results on the shortcoming (3), and it is becoming more popular method for calculations of solid oxides. For the shortcoming (2), a van der Waals functional (vdW – DF) which approximates long-range electron correlation was developed and highly accurate calculations for organic solids and physical adsorption systems etc are now becoming possible.

A number of solid state electronic calculation programs have been developed, and representative codes developed in Japan are shown in Table 5.2. The main difference in these codes is the choice of basis functions and the best basis set depends on your problems you want to calculate. STATE-Senri is a highly tuned program based on pseudopotential and plane wave basis set, and is often used for structures of solids,

Table 5.2 Density functional package available in the world.

Program Name	URL
STATE-Senri	http://www-cp.prec.eng.osaka-u.ac.jp/puki_state/index.php?FrontPage
HiLAPW	http://www.cmp.sanken.osaka-u.ac.jp/~oguchi/HiLAPW/index.html
OpenMX	http://www.openmx-square.org
CONQUEST	http://www.order-n.org

solid surfaces and interfaces, and chemical reaction processes. HiLAPW is a calculation program based on the linearized augmented plane wave (LAPW) method, which is the most accurate program within the density functional method. OpenMX and CONQUEST are the first principle programs of Order N method based on localized basis and are good at calculating large-scale semiconductors and biomolecules.

Program development is prosperous in Japan, and workshops for learning how to use them are also substantial. A workshop called CMD Workshop is held every year. The CMD-workshop covers from the foundation of density functional theory to details of the electronic state calculation method, and even to practical training using the program.[6] Participating in such workshops makes it easy to start the calculation of materials using state-of-the-art first-principles calculation programs.

5.2. First-Principles Simulations on Atomic Geometries and Chemical Reactions at Solid Surfaces and Interfaces

Molecular adsorption and chemical reaction processes on solid surfaces and interfaces are important processes in various fields such as semi-conductor devices and heterogeneous catalysts, fuel cells, secondary batteries, corrosion, and coating. However, it is extremely difficult to elucidate these processes experimentally. For such systems, the first-principle electronic state calculation plays a major role. First-principles electronic state calculation with high prediction accuracy can be said to be an ultrahigh-resolution microscope for both spatial scale and time scale, and it can clarify the atomic/molecular adsorption process and reaction process of the surface and interface which can not be elucidated experimentally. Furthermore, it is expected as a very powerful tool, which gives guidelines for designing more desirable interface in the future. In this section, we introduce an example of the structure and reaction process of the solid surface and interface, which was clarified by first-principles simulation.

5.2.1. *Dependence of molecular adsorption energies on transition metals*

Transition metals are important as catalysts for various heterogeneous catalytic reactions such as ammonia synthesis reaction (Fe), automobile exhaust gas catalyst (Pt, Rh, Pd), methanol synthesis reaction (Cu/ZnO). To clarify factors that govern the reactivity of these metal catalysts is expected to provide guidelines for designing catalysts with higher performance. Molecular adsorption process on transition metal surface is an important process as the first stage of heterogeneous catalytic reaction, so many researches based on experiments and theories have been done for a long time. Experimental results of investigating adsorption energies of various molecules such as carbon monoxide molecules (CO), oxygen molecules (O_2) and hydrogen molecules (H_2) on various transition metal surfaces show that adsorption is generally weak on the right side of the transition metal periodic table, that is, on a metal surface having a large number of d electrons. On the other hand, adsorption is strong on the left side of the periodic table, that is, on a metal surface having a small number of d electrons. This is a common trend for many molecular adsorption. For example, platinum (Pt) is important for CO oxidation catalysts and automobile exhaust gas catalysts, but gold (Au) on the right side of the periodic table is a metal which is very inactive and difficult to adsorb molecules. However, it is known that when it is made into nanoscale clusters, it becomes highly active to act as a CO oxidation catalyst even at low temperature.

For these differences, the interaction between the molecular orbital and the d orbital of the transition metal plays a major role. Figure 5.3 shows the 5σ orbital, the highest occupied molecular orbital (HOMO) of the CO molecule (Fig. 5.3(a)), the $2\pi^*$ orbital which is the lowest unoccupied molecular orbital (LUMO) (Fig. 5.3(b)), the d_{z^2} orbital of

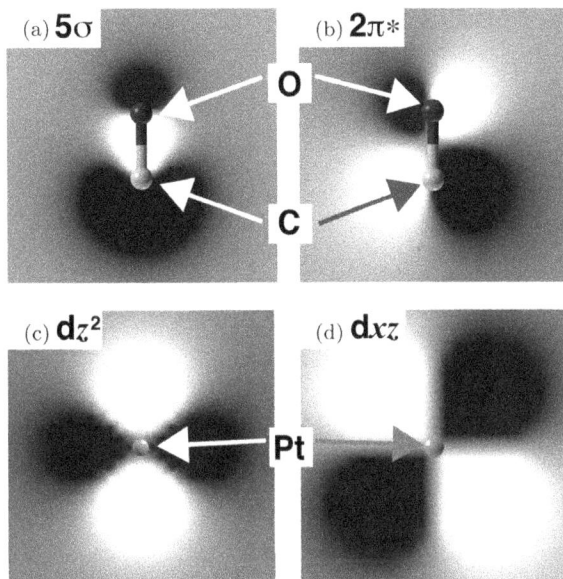

Fig. 5.3 5σ orbital and (b) $2\pi^*$ orbital of CO molecule, (c) d_{z^2} orbital and (d) d_{xz} orbital of Pt.

the Pt atom Fig. 5.3(c)) and the d_{xz} orbital (Fig. 5.3(d)). The white region indicates a region in which the wave function has a positive value, and the black region indicates the region in which the wave function has a negative value. It can be seen that the amplitude of the CO molecule on the C atom side is strong both 5σ and $2\pi^*$. For this reason, it is considered that the CO molecule is more active towards the reaction on the C atom side. In fact, when CO molecules are adsorbed on the Pt (111) surface, at low coverage the C atoms of CO molecules adsorb directly above the surface Pt atoms (called on-top sites). When CO molecules are adsorbed on the on-top site of the Pt surface, the molecular orbital of CO interacts with the d orbitals of the Pt surface atom. The molecular orbitals and the d orbitals have the same symmetry, that is, the 5σ orbital interacts with the d_{z^2} orbital and the $2\pi^*$ orbital interacts with the d_{xz} orbital to form a bonding orbital and an anti-bonding orbital.[7]

In order to estimate the energy generated by the coupling, let us look at the case of diatomic molecules as shown in Fig. 5.4. Let the orbital

Atom A   Molecule AB   Atom B

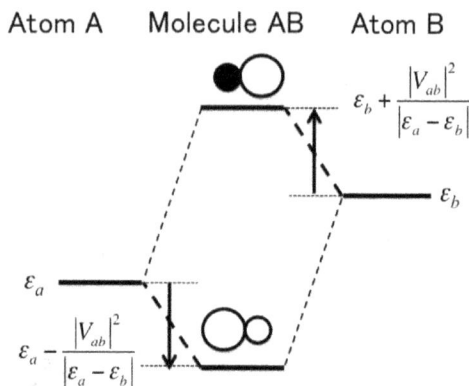

Fig. 5.4   Change in energy levels due to the bond formation between atom A and atom B. Bonding and anti-bonding orbitals are also shown schematically.

energies of isolated atoms A and B be ε_a and ε_b, respectively. As atoms approach each other, they begin to interact, resulting in overlapping wave functions. Then, the Hamiltonian matrix elements $V_{ab} = \langle \psi_a | \hat{\mathcal{H}} | \psi_b \rangle$ become non-zero. Then, a change occurs in the energy levels as shown in Fig. 5.4, the lower energy level ε_a is even lower, $\varepsilon_a - |V_{ab}|^2/|\varepsilon_a - \varepsilon_b|$, and the energy level on the higher side ε_b is even higher $\varepsilon_b + |V_{ab}|^2/|\varepsilon_a - \varepsilon_b|$. In addition, the molecular orbital generated by hybridization is also schematically shown in Fig. 5.4. Since the energy level of the bonding orbital is close to the atom A, the proportion of the atom A's orbital is large and the proportion of the atom B is small. Conversely, since the energy level of the anti-bonding orbital is closer to the atom B, the proportion of the atom B is larger and the proportion of the atom A is smaller.

The molecular orbital of CO and the d orbital of Pt basically have similar orbital hybridization, which is called the Blyholder mechanism. As shown in Fig. 5.5, the interaction between 5σ which is the HOMO and the d orbital is called σ donation, and partial electron transfer occurs from the occupied 5σ to the empty d orbital. On the other hand, the interaction between $2\pi^*$ which is the LUMO and the d orbital is called π back-donation, and partial electron transfer occurs from d orbital to $2\pi^*$ which was the vacant orbital. These electron transfer causes a bond to be formed between the CO molecule and the transition metal, but

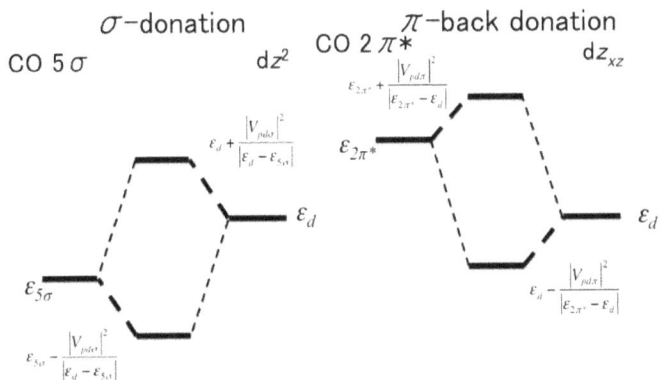

Fig. 5.5 Interaction between CO molecule and d orbitals of transition metal (Blyholder mechanism). The left panel shows the hybridization between 5 σ orbital which is the HOMO of CO and d orbital. Electrons partially move from 5 σ, which was originally fully occupied, to the d orbital which was partially unoccupied (σ donation). The right panel shows a hybridization between CO 2 π^* orbital (LUMO) and d orbital of a transition metal. Electrons partially move from the occupied d orbital to the empty π orbital (π back-donation).

with respect to the C–O bond in the molecule, since π^* which is an anti-bonding orbital, is partially occupied, it becomes weak.

Now, we know the adsorption mechanism of CO, we will investigate the difference in adsorption energy between Pt surface and Au surface. The d orbital on the surface of Pt and Au forms a complicated band structure and the energy is distributed with a width. The situation is shown in Fig. 5.6. Looking at this figure, we can see that d orbital is almost occupied for both Pt and Au. Then, both the 5σ orbital and the d orbital are occupied and therefore, the anti-bonding orbital of the hybridized state is also occupied, and therefore, on the surface of Pt or Au the donation contributes little to the adsorption energy of CO, rather it is repulsive. On the other hand, in the interaction between $2\pi^*$ and the d orbital, only the coupling trajectory is occupied, and a stable coupling is created. Therefore, when CO molecules are adsorbed on the metal surface on the right side of the transition metal periodic table such as Pt and Au, it can be said that stabilization of energy by π back-donation dominates binding. According to Fig. 5.5, it can be seen that when $V_{pd\pi}$ is large or when the d orbital is close to $2\pi^*$ orbital, ie, shallower, the stabilization by the interaction energy is larger. When comparing Pt and

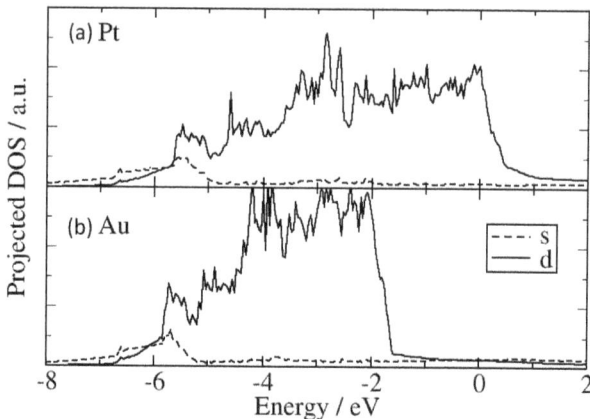

Fig. 5.6   Density of states of the Pt (111) surface and the Au (111) surface. Density of states projected on the s orbital and d orbitals are indicated by broken lines and practice, respectively. Zero of energy is the Fermi level.

Au with respect to $V_{pd\pi}$, the d orbital of Pt spreads more because it has less atomic nucleus charge than Au, so $V_{pd\pi}$ is larger for Pt, stabilizing the binding energy. Furthermore, as can be seen from Fig. 5.6, the d band of Pt is shallower than the d band of Au. This also contributes to stabilizing the binding energy of CO on Pt compared with Au. When the periodic table goes further to the left from Pt, this tendency becomes stronger and the stabilization by adsorption energy becomes larger. Then, why does catalytic activity increase when gold turns into nanoclusters? This is because gold at the edge and vertex of the cluster has fewer adjacent gold atoms, that is, coordination number. The d orbital of such a gold atom approaches the Fermi level as the bandwidth becomes narrower. Therefore, it is considered that catalytic reactivity comes out to the same degree as Pt.[8]

5.2.2.   *Electrochemical reactions between water/Pt interfaces*

In the previous section, it was shown to clarify factors that govern the reactivity of the local structure of metal surfaces and clusters by calculating electronic states. This is a major guideline in designing a new catalyst. On the other hand, complicated catalytic reactions need be understood by clarifying their reaction paths. Experimentally it is

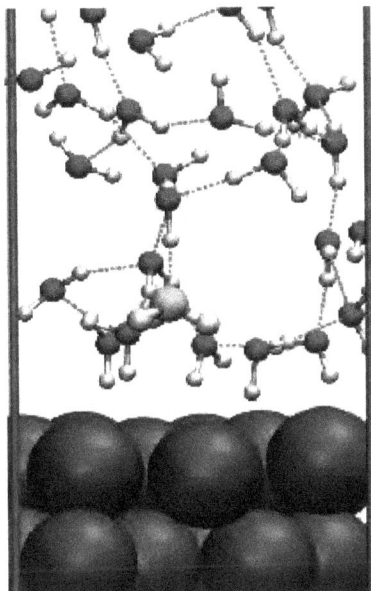

Fig. 5.7   Hydrogen evolution reaction at water/Pt (111) interface. Volmer process in which one of protons in a hydronium ion in aqueous solution adsorbs onto Pt surface. Reprinted from Ref. [9].

difficult to investigate the chemical reaction process in many cases, and the expected role of the first-principles electronic state calculation is extremely large. Elucidating the chemical reaction process at the solid-liquid interface like the electrochemical reaction is experimentally very challenging task, but it was elucidated before the experiment by the first-principles electronic state calculation simulation. Figure 5.7 shows the process of so-called Volmer process in which hydronium ions in aqueous solution receive electrons from Pt electrode and become adsorbed hydrogen at the interface with water and Pt (111) in the first-principles molecular dynamics simulation.[9] It can be said that this is the first successful example in the world to see the instant at which an electrochemical reaction occurs by controlling the electrode potential.

5.2.3.   *Simulations of etching processes at Pt/HF/SiC interfaces*

In the previous section, the reaction process at the solid-liquid interface was shown, but the reaction at the interface where the solid/liquid/solid

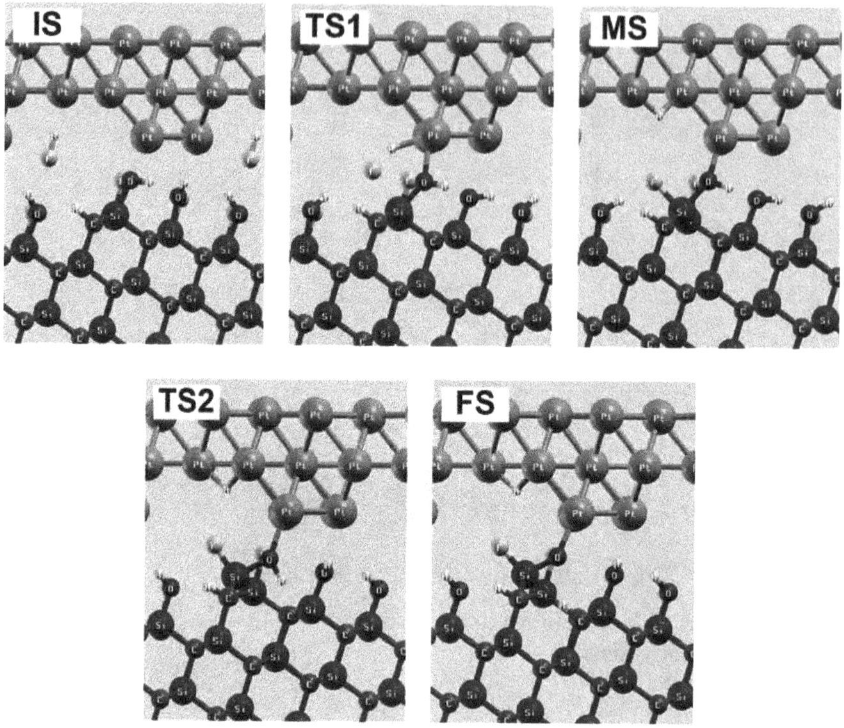

Fig. 5.8 Dissociation reaction process of HF molecules at Pt/HF/SiC (001) interface. The overall activation barrier is lowered by stabilizing the penta-coordinated Si by forming a bond with Pt in the meta-stable reaction intermediate. Reprinted Ref. [10].

three materials are in contact is important in the field of surface processing including polishing, and the fuel cell electrocatalysts. Although it is an important reaction process in application, it is currently very difficult to elucidate the reaction process experimentally. For such systems accurate and efficient first-principles theoretical simulations play important roles. Figure 5.8 shows the initial reaction process of HF solution etching of SiC surface by Pt catalyst.[10] SiC is promising as a next generation power device material together with GaN. Although it is indispensable to create a flat surface to create a high performance device, it is difficult to create a flat surface at the atomic level because it is mechanically hard and chemically inert semiconductor. It was noted that Osaka University group was able to obtain a flat surface at atomic level by polishing SiC surface with HF aqueous solution using Pt catalyst.

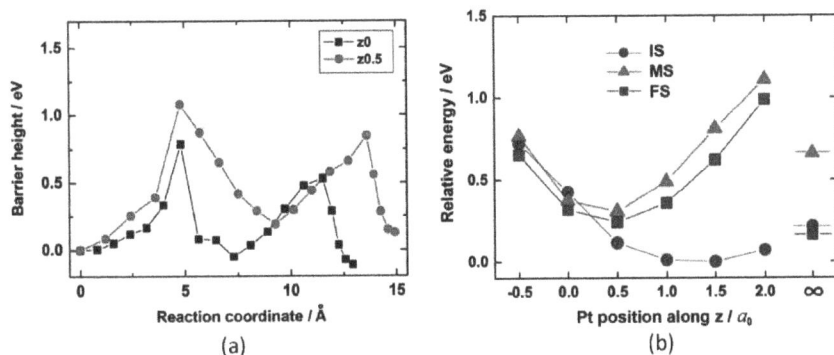

Fig. 5.9 (a) Energy profile of HF dissociation reaction at Pt/HF/SiC interface. (b) Relative energies of the initial (IS), meta-stable (MS), and final (FS) states of the reaction as a function of Pt-SiC distance. Reprinted from Ref. [10].

By the first-principles simulation, the initial process of the reaction process was clarified.

As shown in Fig. 5.8, in the early stage of the etching reaction, HF molecules are dissociated and adsorbed. It was found that fluorine ions were bonded to the surface first layer Si to form pentacoordinated Si as a reaction intermediate (MS). Furthermore, it was also revealed that the stability of this MS largely depends on the relative position of Pt and SiC, especially the distance between SiC and Pt. Fig. 5.9(a) shows the energy profile of the reaction process. It can be seen that the activation energy of dissociation reaction of HF greatly changes depending on the distance between Pt and SiC. Figure 5.9(b) shows the Pt-SiC distance dependency of the relative stability of the initial state (IS), the MS, and the final state (FS). When Pt approaches, IS becomes unstable, whereas activation barrier lowers because MS and FS are stabilized. It was also revealed that this stabilization is due to the bond formation between Pt and the adsorbed oxygen atom.

5.2.4. *Simulations of GaN crystal growth under high pressure and high temperature*

Although chemical reactions under high temperature and high pressure are extremely important for applications, it is often difficult to elucidate the reaction mechanisms. In such systems, first-principles simulations with high reliability and wide application range play important roles.

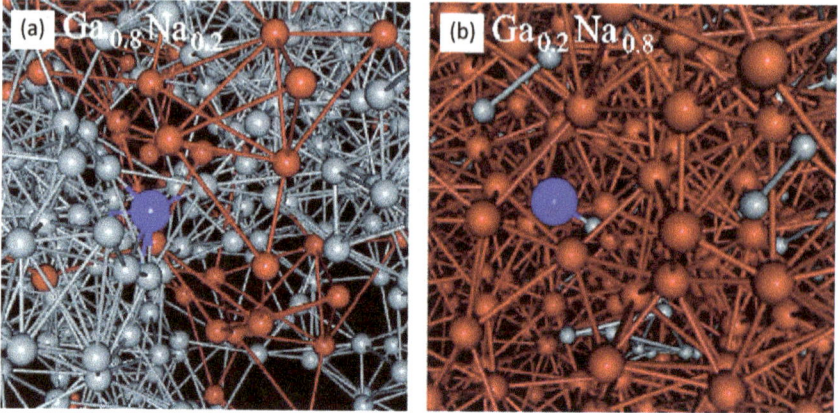

Fig. 5.10   (a) Dissolution of a N atom in Ga rich GaNa liquid alloy, (b) Dissolution of a N atom in Na rich GaNa liquid alloy. Reprinted from Ref. [11].

GaN is an important material for optoelectronic devices and also promising material for power devices. It is necessary to grow crystal with lower concentration of defects in order to produce a device with higher performance. The Na flux method has been proposed as a promising method for growing bulk GaN crystal as a substrate for homoepitaxial growth. To grow GaN crystal by dissolving N atoms in Ga metal, very high temperature and pressure conditions such as 1800–2300 K and 1000–1500 MPa are required because the solubility of N into pure Ga metal is very low. However, by alloying Ga and Na, the solubility of N atoms dramatically increases, and the required temperature and pressure can be extremely lowered to 1000–1100 K, 3–5 MPa, respectively. In particular, the increase in the solubility of N atoms becomes prominent in Na-rich GaNa alloy. Since the N atom solubility in pure Na is as low as the nitrogen solubility in pure Ga, the solubility of N atom increases by alloying of Ga and Na, but its microscopic mechanism is not known because it is difficult to clarify experimentally. Therefore, we have performed first-principles molecular dynamics simulations of N dissolved into Ga-Na liquid alloys. Figure 5.10 shows the result of simulation of N atom dissolved in Ga–Na alloy liquid at 1073 K by the first-principles molecular dynamics method.[11] As shown in Fig. 5.10(a), in a Ga-rich GaNa alloy, a metallic network of Ga is formed, Ga is stabilized, and N atoms are weakly adsorbed to the interface between

Ga and Na. On the other hand, as shown in Fig. 5.10(b), Ga forms small clusters in the Na-rich GaNa alloy, they become chemically very reactive and they bind N atoms quite strongly. This is a physical factor that dramatically increases the solubility of N atoms in Na-rich GaNa alloys.

Another interesting thing is that Na flux method dramatically promotes single crystal growth of GaN when carbon is added. The mechanism of carbon additive to supress the poly-crystal growth and enhance the single crystal growth was recently theoretically clarified and reported.[12] As shown in Fig. 5.11, in the Na flux, carbon atoms bind with nitrogen atoms to form stable CN ions, thereby suppressing the formation of microcrystals, whereas when they reach the interface with the GaN single crystal, the CN bonds are easily dissociated to promote the growth of single crystal of GaN.

5.3. Molecular Simulation for Biomolecules

5.3.1. *Biochemical functions of biomolecules*

A living thing consists of biomolecules and is a complex system in which these molecules dynamically interact with each other and yield the biological functions. In particular, proteins show prominent functions such as high-specific molecular recognitions and play essential roles in the life system. Proteins are usually built from a repertoire of 20 amino acids, each with different chemical properties and molecular structures such as size, shape, charge, hydrogen-bonding capacity, hydrophobic character. Various functions of proteins are attributed to the versatility of these 20 building blocks. For example, photosystem II (PSII), which catalyzes the oxidation of water to molecular oxygen at the Mn cluster at room temperature, serves as a high-efficient catalyst. Heme proteins are involved in various biochemical functions, including oxygen transportation, electron transfer, and oxygenation or oxidation of organic molecules, though their active sites contain the same cofactor, heme. This indicates that heme active sites are multifunctional. Elucidation of the mechanisms of biochemical functions of proteins leads to design the new functional molecules as well as provides insights into biology. What should we elucidate to understand the mechanisms? A protein forms an appropriate shape for the function, and changes the shape to exert its function. It is,

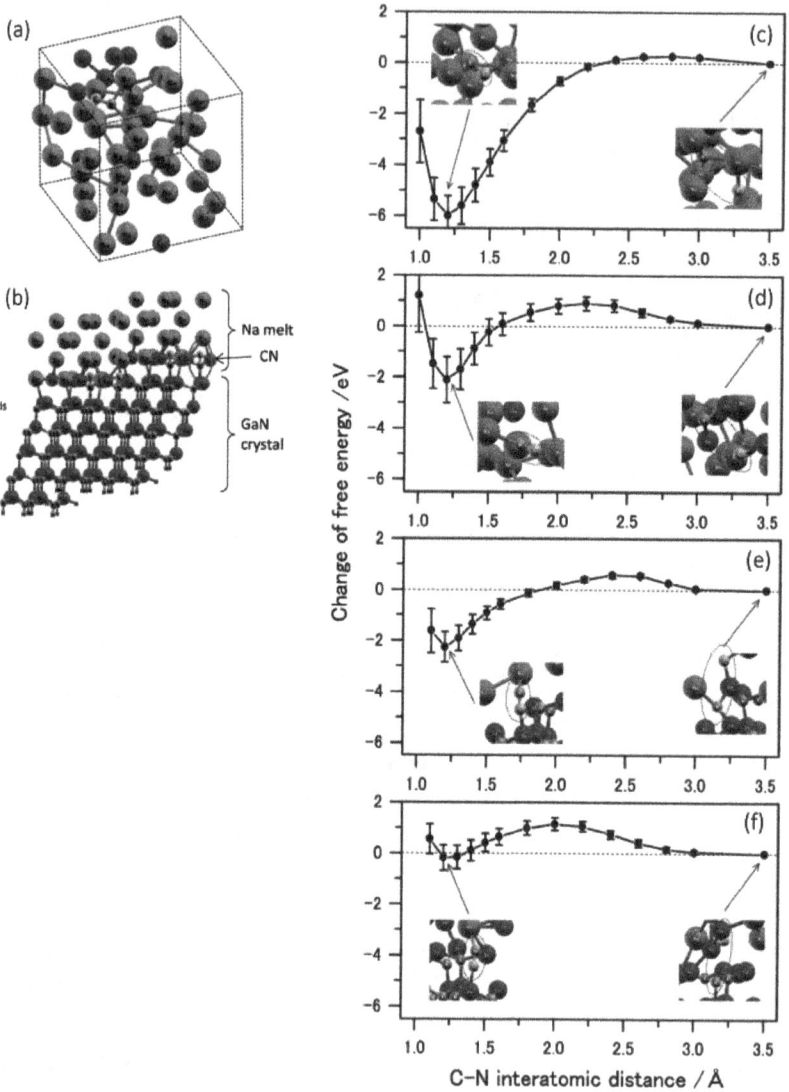

Fig. 5.11   (a) Structure of C and N atoms in Na flux. (b) Interface between Na flux and GaN single crystal. (c)–(f) Dissociation free energy profile of C–N bond. (c) in Na flux (d) in Na flux in which Ga exists (e), (f) C–N bond dissociation at GaN single crystal interface. Reprinted from Ref. [12].

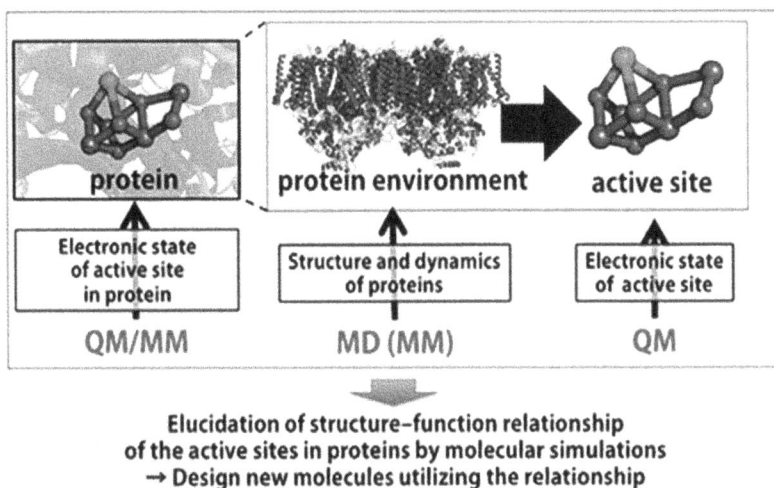

Fig. 5.12 Objective of molecular simulation for biomolecules.

therefore, essential to elucidate the molecular and electronic structures and their dynamics of the active site in a protein in order to understand the mechanism of the function of a protein (Fig. 5.12).

Molecular simulations have increasingly become useful in studying proteins, because it is possible for molecular simulations to obtain their transient structures when the proteins exert the functions.[4] Here, we introduce electronic structure calculations (molecular orbital method), molecular dynamics (MD) simulations, and QM/MM (Quantum Mechanics/Molecular Mechanics) calculations, which have recently used for protein sciences. In the next subsection, we explain the application of the electronic structure calculations to heme, which is often found in the active site of proteins. Next, we introduce the applications of the MD simulations and the QM/MM calculations.

5.3.2. *Electronic structure calculations for biomolecules (Molecular orbital method)*

We introduce the study of the relationship between heme distortion and redox potentials of heme as an application of the electronic structure calculations described in the Subsection 5.1. Heme is a widely-distributed cofactor in biology, and consists of an iron ion and a porphyrin molecule.

Fig. 5.13   Normal-coordinate structural decomposition of heme distortion.

Heme protein, a metalloprotein comprising heme groups, serves electron transfer, oxygen transport, and redox catalysis. Recently, it has been suggested that heme distortion also controls the properties of heme such as the redox potential and the ligand binding affinity. Normal-coordinate structural decomposition (NSD),[13] which converts a complex structural deformation into a summation of simple symmetric deformations (Fig. 5.13), has been used to quantify the structural deformation of heme and to investigate the relationships between biological functions and heme conformations. Here, we performed DFT calculations on distorted heme models with the PBE0 exchange–correlation functional and examined the relationship between heme distortion and redox potential. We focused on the ruffling saddling, doming, and breathing distortions of heme, which are often found in protein data bank (PDB). Our computation showed that the ruffling deformation basically decreased the redox potential. In contrast, the saddling, doming, and breathing deformations monotonically increased the redox potential An in-plane mode (breathing mode) affected the redox potential rather than out-of-plane modes (ruffling, saddling, and doming modes) (Fig. 5.14).

5.3.3.   *Molecular dynamics (MD) simulation for biomolecules*

Since proteins are flexible, the structural change is often crucial to protein functions. Hence, understanding how proteins function requires capturing the transient structures involved in protein functions. MD simulations are appropriate for obtaining the transient protein structure and observing essential protein motions.

An MD simulation numerically solves the Newton's equations of motion as shown in Eq. (5.12), and computationally mimics the motion of a system composed of a number of particles.

$$m_i \frac{d^2 r_i}{dt^2} = -\nabla_i U(\mathbf{r}_1, \mathbf{r}_2, \ldots, \mathbf{r}_N) \quad i = 1 \ldots N, \tag{5.12}$$

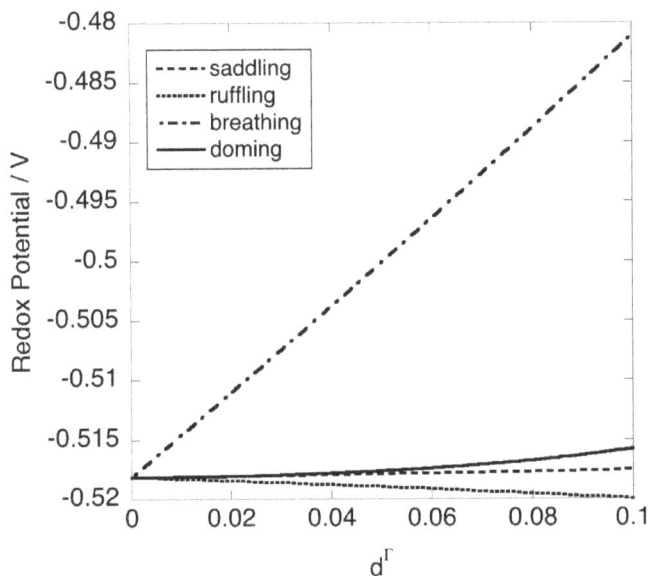

Fig. 5.14   Relationships between heme distortion and redox potential of heme.

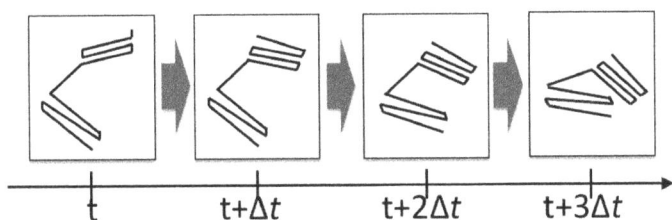

Fig. 5.15   Schematic view of an MD simulation.

where m_i and $\mathbf{r}_i$ denote the mass and position vector of the i-th particle, and $U(\mathbf{r}_1, \mathbf{r}_2, \ldots, \mathbf{r}_N)$ expresses the potential energy of the system. In an MD simulation of a protein, we regard atoms composed of the system as particles. Solving the Newton's equation of motion with the difference method such as the Verlet method, the velocity Verlet method, and the leap-frog method, we obtain new positions of particles at $t + \Delta t$ from the positions at t. Repeating it, we can simulate time-evolution of the system. Finally a trajectory of the system is obtained as shown in Fig. 5.15. The potential energy was expressed as

$$U = U_{\text{bond}} + U_{\text{angle}} + U_{\text{torsion}} + U_{\text{improper\_torsion}} + U_{\text{electrostatic}} + U_{\text{vdW}},$$

$$(5.13)$$

where U_{bond} and U_{angle} denote the bond stretching and angle bending terms, based on the Hook's law. U_{torsion} and $U_{\text{improper\_torsion}}$ describe the dihedral torsion and improper torsion terms. These four terms are called bonded terms because these are involved in covalent bonds. $U_{\text{electrostatic}}$ denotes the electrostatic term and represents the electrostatic interactions between two atoms. U_{vdW} denotes the van der Waals term and is usually expressed with the Lennard–Jones potential. The functional form and the parameter sets to calculate the potential energy are called the force field.

Since the transitions of protein structures are strongly related to their functions, the sampling of protein structures enables us to understand the biochemical functions. However, the structural sampling still remains as a challenging problem in computational protein sciences due to the accessible time-scales of conventional MD simulations. The conformational transitions relevant to biological functions usually occurs on a timescale ranging from μs to ms, although conventional MD simulations using an all-atom model follow timescales of ns. To overcome the time-scale limitation of the conventional MD simulations, many enhanced conformational sampling methods have been developed. Here, a simple but efficient protein conformational sampling method, called the TaBoo SeArch (TBSA) algorithm, was developed.[14] In TBSA, the rarely appealing structures are intensively re-sampled. Then these structures are selected as "seeds" for the conformational re-sampling based on an inverse histogram on the energy space, leading to the enhancement of the conformational sampling of proteins. TBSA consists of repeats of the following steps: (i) Selection of seeds based on the inverse histogram and (ii) Conformational re-sampling from the selected seeds by short-time MD simulations, where the conformational re-sampling from the selected seeds is performed independently and simultaneously (Fig. 5.16). TBSA is, thus, remarkably easy to use without modifications of the source codes of MD programs. As a demonstration, TBSA was applied to the folding of chignolin, and automatically sampled

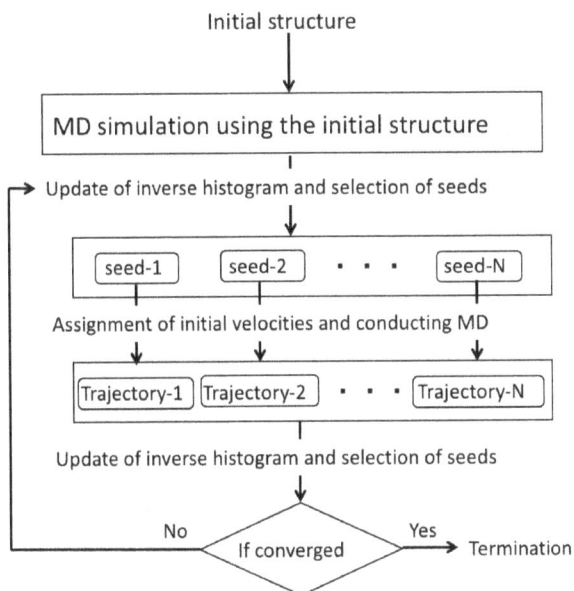

Fig. 5.16   Algorithm of TBSA.

Table 5.3   Popular software packages for molecular dynamics simulations and their website.

Software packages	Website
AMBER	http://ambermd.org/
CHARMM	http://www.charmm.org/
GROMACS	http://www.gromacs.org/
NAMD	http://www.ks.uiuc.edu/Research/namd/
myPresto	http://presto.protein.osaka-u.ac.jp/myPresto4/

the native structure (root mean square deviation <1.0 Å) from an extended structure with ns-order computational costs started, although a long-time 1-μs normal MD simulation failed to find the native structure.

Table 5.3 shows popular software packages of MD simulations. AMBER (Assisted Model Building with Energy Refinement) is developed by P. A. Kollman and co-workers, and is one of the most popular program suites and force fields for MD simulations. It also has a user-friendly interface. CHARMM (Chemistry at HARvard Macromolecular Mechanics) is developed by M. Karplus and co-workers. They also develop the force fields called CHARMM force fields, which are appropriate for the MD simulations of membrane proteins. GROMACS (GROningen MAchine for Chemical Simulation) is suited for parallelization on processor clusters and a very fast program for MD simulations. NAMD (Nanoscale Molecular Dynamics program) is a parallel molecular dynamics code designed for high-performance simulation of large biomolecular systems. myPresto (Medicinally Yielding Protein Engineering SimulaTOr) is developed by H. Nakamura and co-workers in Japan. A site license for myPresto is available at no cost to both academic and industrial users.

As an example of studying with MD simulations, we introduce the study of facilitation of ligand access to the active site in monoamine oxidase A (MAOA) by membrane attachment. MAOA is a membrane protein with a single transmembrane helix and catalyzes the oxidative deamination of extra- and intracellular transmitters. It is, therefore, one of the major target for development of central nervous system drugs. Although the X-ray structures of MAOA were determined at high resolution, the route for ligand access to the active site remains unclear. MD simulations on MAOA were performed to find the route. Our MD simulations showed that interdomain anticorrelated movements in the membrane-bound system induced structural change, resulting in the opening pathways for substrate uptake and product release, as shown in Fig. 5.17.[15]

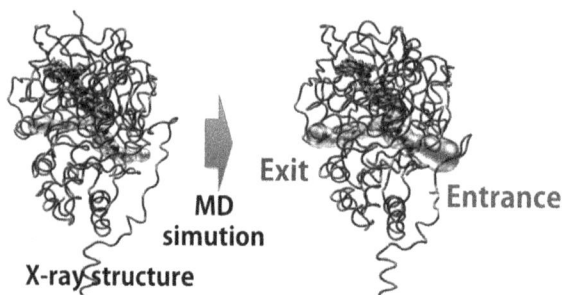

Fig. 5.17   A route for ligand access to the active site of MAOA.

5.3.4. *QM/MM simulation for biomolecules*

In general, proteins function at the local region called the active site, where a catalytic chemical reaction occurs in an enzyme, an electron is received and released in an electron transport protein, or a chromophore is excited in a photoactive protein. Since chemical reactions and electron transfer are quantum-mechanical phenomena, active sites should be treated in a quantum-mechanical manner. In addition, the dynamical characteristics of proteins are also regarded as intrinsic fundamental properties of their functions, implying that the entire systems, including proteins and solvent molecules, should be considered in an investigation of the protein dynamics. Therefore, massive computer resources are required for both the quantum-mechanical and dynamical analyses of proteins. Hybrid quantum mechanical/molecular mechanical (QM/MM) calculations are applicable for the assessment of biomolecular systems. These methods have become powerful tools for investigating biochemical reactions in proteins. In such simulations, the central reactive region is treated quantum mechanically, to allow key bonds to form and break, whereas the rest of the macromolecule as well as some explicit solvent molecules are modeled by MM methods, to make the calculations computationally feasible. The total energy of the whole system, E_{tot}, is represented as

$$E_{tot} = E_{QM} + E_{MM} + E_{QM/MM}, \qquad (5.14)$$

where the E_{QM} and E_{MM} terms denote the energies of the QM and MM parts, respectively. The $E_{QM/MM}$ term includes the interaction energies between the QM and MM parts, and is composed of the electrostatic interaction, E_{ele}, and the van der Waals interaction, E_{vdW}, as

$$E_{QM/MM} = E_{ele} + E_{vdW}. \qquad (5.15)$$

In the QM part, Hartree–Fock (HF), density functional theory (DFT), configuration interaction singles (CIS) and doubles (CIS(D)), second-order Møller–Plesset perturbation theory (MP2), complete active-space configuration interaction (CASCI), and complete active-space self-consistent field (CASSCF) are used. The AMBER force fields and the CHARMM force fields can be adopted in the MM part.

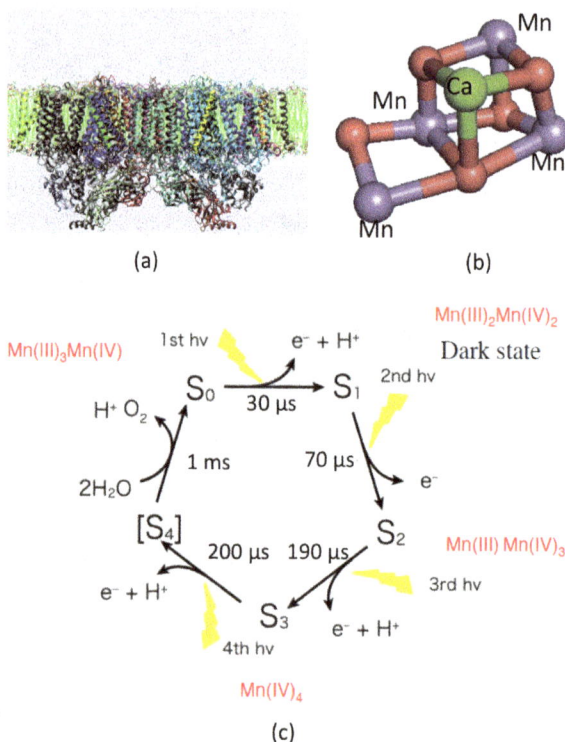

(a)

(b)

(c)

Fig. 5.18 Cartoon structure of photosystem II (PSII) (a), Oxygen evolving complex of PSII (b), and Kok cycle (c).

For the QM/MM boundary problem, the link atom method is often implemented in the QM/MM code. A hydrogen atom was put on the line that connected from QM boundary to MM boundary atoms. To avoid the additional unnatural degree of freedom in QM/MM system, the length between the QM boundary atom and the hydrogen atom set to the fixed value. The hydrogen atom was then treated as a QM atom in the QM region. The MM boundary atom and the atoms chemically bounded to the MM boundary atoms dose not electrostatically interact with the atoms in the QM regions. The omitted charge was redistributed to corresponding residue in the atoms of the MM regions. If at least one QM atom was involved in two, three, and four body bonded MM interactions from the MM system, the bounded MM interactions were applied to the QM atom.

As an application of the QM/MM calculations, we introduce the study of the electronic structure of the Mn cluster in the photosystem II of photosynthesis. Photosynthesis is a process to convert light energy into chemical energy. The Mn cluster in the PSII catalyzes the oxidation of water yielding oxygen molecules (Fig. 5.18(a) and (b)). In 2011, the X-ray structure of the PSII was reported at 1.9 Å resolution, and the detailed structure of the Mn cluster was first determined.[16] However, the structure of the Mn cluster was only reported in the dark-stable S_1 state of the Kok cycle (Fig. 5.18(c)), and those in the other states (S_0 and S_2–S_4 states) were not determined. In addition, the electronic structures of the Mn cluster were unclear. The QM/MM calculations[17] were performed on the Mn clusters determined with recent X-ray crystallography. The valences of each manganese atom were assigned as $Mn1^{III}$, $Mn2^{IV}$, $Mn3^{III}$, and $Mn4^{III}$ in the S_0 state, $Mn1^{III}$, $Mn2^{IV}$, $Mn3^{IV}$, and $Mn4^{III}$ in the S_1 state, and $Mn1^{III}$, $Mn2^{IV}$, $Mn3^{IV}$, and $Mn4^{IV}$ in the S_2 state. The computational assignment in the S_1 state corresponds to that based on the X-ray free electron laser (XFEL) structure.[18]

References

[1] P. Atkins and R. Friedman, *Molecular Quantum Mechanics Fifth Ed.*, Oxford, (2011).
[2] A. Szabo and N. S. Ostlund, *Modern Quantum Chemistry*, Dover (1989).
[3] D. B. Cook. *Handbook of Computational Quantum Chemistry*, Dover, (2005).
[4] A. R. Leach, *Molecular Modelling*, Prentice Hall, (2001).
[5] R. M. Martin, *Electronic Structure: Basic Theory and Practical Methods*, Cambridge Univ. Press, (2008).
[6] CMD Workshop: http://phoenix.mp.es.osaka-u.ac.jp/CMD/
[7] B. Hammer, Y. Morikawa, and J.K. Nørskov, *Phys. Rev. Lett.* **76**, 2141 (1996).
[8] K. Okazaki, *et al.*, *Phys. Rev.* **B69**, 235404 (2004).
[9] M. Otani, *et al.*, *J. Phys. Soc. Jpn.* **77**, 024802 (2007).
[10] P. V. Bui, *et al.*, *Appl. Phys. Lett.* **107**, 201601 (2015).
[11] M. Kawahara, *et al.*, *J. Appl. Phys.* **101**, 066106 (2007).
[12] T. Kawamura, *et al.*, *App. Phys. Express* **9**, 015601 (2016).
[13] W. Jentzen, X.-Z. Song, and J. A. Shelnutt, *J. Phys. Chem. B* **101**, 1684 (1997).
[14] R. Harada, Y. Takano, and Y. Shigeta, *J. Comput. Chem.* **36**, 763 (2015).
[15] R. Apostolov *et al.*, *Biochemistry* **48**, 5864 (2009).
[16] Y. Umena *et al.*, *Nature* **473**, 55 (2011).
[17] M. Shoji *et al.*, *Catal. Sci. Technol.*, **3**, 1831–1848 (2013).
[18] M. Suga *et al.*, *Nature*, **517**, 99–103 (2015).

Chapter 6

THEORY OF ELECTRICAL CONDUCTION

Nobuhiko Kobayashi
University of Tsukuba

Hiroyuki Ishii
University of Tsukuba

Kenji Hirose
NEC Corporation

In this chapter, we describe the theory of electrical conduction, especially the analysis of carrier transport in organic semiconductors.
Organic semiconductors are crystals that are assemblies of π-conjugated
molecules weakly bonded by van der Waals interactions. For the analysis
and control of the carrier transport characteristics of organic semiconductors, it is indispensable to elucidate and control the scattering
mechanism at defects, the thermal fluctuation, and the interface with a
substrate. We introduce research on electron and hole carrier transport
in soft crystals by analysis of the transport characteristics on the basis of
first-principles calculation.

6.1. Theory of Electrical Conduction in Semiconductor Devices and Problems

Semiconductor technology is a fundamental technology supporting our modern information society. Semiconductor devices such as transistors are incorporated in various electronic devices, which have been developed through the progress of miniaturization and integration techniques for hard inorganic materials, principally silicon. However, the microfabrication and integration technologies are approaching their limitations, and increasing research is being performed on devices made of various semiconductor materials. For example, ultrahigh-speed transistors made of carbon nanotubes or graphene, transistors with ultralow power consumption made of magnetic semiconductors, and flexible transistors made of organic semiconductors, which is the main subject in this chapter, have attracted much attention as new electronic devices.

For the development and application of these devices, research by performing transport simulations of carriers such as electrons and holes, spins, and heat flowing in the devices is indispensable. The Boltzmann equation method is a conventional method widely used in charge transport simulations, where collision processes of electrons are treated by considering the probability distribution of electrons using probability theory. However, its effectiveness is uncertain in the micro scale region toward the atomistic level, which the current device size is approaching.

On the other hand, at the surface and interface of semiconductors, which have a significant effect on device performance, the process technology for forming devices has already been controlled at the atomistic level, and large-scale calculations called first-principles electronic states calculations, which precisely consider chemical bonds between atoms and the electron density on the basis of quantum mechanics, have been very successful in the theoretical analysis of atomistic structures. In these calculations, groups of electronic states called bands are obtained, and metals, semiconductors, and insulators are distinguished by the energy band structures. Even the effects of micro scale surface and interface structures can be clarified from their influence on the band structure. Furthermore, on the basis of static electronic states calculations, first-principles calculation methods for dynamic charge transport simulations have also been developed by precisely calculating the electron density from chemical bonds between atoms.[1-3] Using such methods, it is possible to investigate the charge transport properties of device structures consisting of not only conventional silicon materials but also organic materials, molecular materials, and various other new materials. Furthermore, it is also possible to analyze quantum effects such as discrete energy levels, tunneling effects, and spins, which are noticeable at the nanoscale or near-atomistic level, and to calculate collisions between electrons accurately.

We describe difficult points in constructing this new calculation method for electrical conduction. In conventional static first principles calculations which accurately treat the electron density distribution of the surface and the interface of a semiconductor, as previously described, the electronic states are calculated using standing waves for a periodic structure consisting of repeating unit cells. On the other hand, in a dynamic phenomenon in which carriers move, such as electrical conduction, the electronic states are described as scattering waves. Scattering waves move as a result of an electric field, temperature difference, and so forth, and are repeatedly scattered by impurities and lattice vibrations in the device, resulting in resistance and determining the device characteristics. The difficulty lies in converting standing waves into scattering waves and calculating the scattering processes. In the Boltzmann equation method, the effect of these scattering processes is stochastically calculated without considering the nature of the wave of electrons, but in the first-principles calculation method, the electronic states described by these scattering waves are precisely determined on the basis of quantum mechanics. This approach is a challenging problem in the field of computational physics since a large amount of computational time and high computational stability are needed.

Next, we describe how the electrical transport properties change as a device becomes smaller. When a device is sufficiently large, electrons are repeatedly scattered by impurities, lattice vibrations and electrons, which cause resistance. This is called the diffusive regime. However, when a device becomes extremely small, the scattering becomes rare, and furthermore, electrons significantly exhibit wave nature and pass without collision. Electrons are scattered at the edge of the device structure and the resistance depends only on the device shape. Such a regime is called the ballistic regime. Semiconductor devices emit large amounts of heat during operation. Some semiconductor devices also emit light. Heat and light emission appear as energy emission in processes in which electrons repeatedly undergo scattering, but it is extremely difficult to understand exactly what processes occur and what mechanisms cause heat and light emission. In the Boltzmann equation method, each process is stochastically incorporated, but the investigation of these phenomena on the basis of quantum theory is also an important issue for first-principles calculation.

We outline the conductivity of a semiconductor device using flexible organic materials. The movement of carriers such as electrons and holes in a soft organic material has similarities and differences from conduction in a hard crystal structure as described above. Similar points are that carriers have wave properties and that scattering due to impurities and the lattice vibration determines device characteristics for highly crystalline structures. On the other hand, a difference is that when the crystal structure is markedly deformed owing to its softness, the carriers cannot move as a wave and hop probabilistically as particles in the classical regime. This point can be clearly observed in the temperature dependence of device characteristics. In the case where the carriers have the properties of waves, as the temperature increases, the lattice vibration is activated and the frequency of scattering increases, so the conductivity decreases. This is a common phenomenon, even for hard materials such as silicon and is called band transport. On the other hand, when the carriers have the properties of particles, as the temperature increases, they move upon receiving thermal energy and the conductivity increases. This is a phenomenon seen in organic materials with very low conductivity, which is called thermally activated hopping transport. In recently developed organic semiconductors, the crystallinity has been greatly improved by single-crystallization, and owing to the coexistence and competition between properties of individual constituent molecules and solid states with periodic crystalline structures, characteristic features with intermolecular carrier transfer caused by both hopping and band transport mechanisms appear.

Electronic devices are currently composed of various semiconductor materials and we have explained that the electrical conduction characteristics depend on the individual materials, their sizes, and structures. Moreover, for theoretical electrical conductivity analysis, a calculation method based on first-principles theory becomes indispensable as an alternative to the conventional Boltzmann equation method as the device approaches the atomistic level. In the following sections of this chapter, we will describe the electrical conduction phenomena in single-crystal organic semiconductor materials, which have recently attracted much attention in the field of semiconductor devices.

6.2. Organic Semiconductor

An organic semiconductor device is an electronic device made of an organic material with the properties of a semiconductor. Organic semiconductors are soft and lightweight, making it possible to realize curved displays and devices that can be attached to the human body, which cannot be achieved with conventional electronic devices. Organic semiconductors are crystals consisting of π-conjugate covalent molecules weakly bonded by van der Waals interactions. Conventionally, such condensed matters does not have good crystallinity or conductivity. However, single-crystallization has recently been achieved, markedly improving the conductivity. Furthermore, in contrast to crystallization by a high-temperature melting process for conventional inorganic silicon crystals, organic semiconductors can be fabricated by a low-cost solution process at room temperature. Thus, organic semiconductors are highly promising as next-generation electronic device materials. Typical single-crystal organic semiconductors with high mobility include rubrene and pentacene, and organic semiconductor transistors using these materials have been developed. It is now expected that the mechanism of carrier transport in organic semiconductors can be elucidated and that materials exhibiting high mobility can be developed by novel molecular synthesis.[4] For the analysis and control of the carrier transport characteristics of organic semiconductors, it is indispensable to elucidate and control the scattering mechanism at defects, the disorder, the thermal fluctuation, the interface between the substrate and electrode, and so forth, in addition to the properties of individual constituent molecules and the properties of their solid states. For this reason, it is important to carry out detailed crystal structure analysis at the atomistic level by combining theory and experiments, and carrier transport analysis taking into consideration various scattering mechanisms.

Pentacene

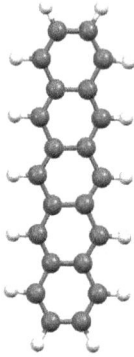
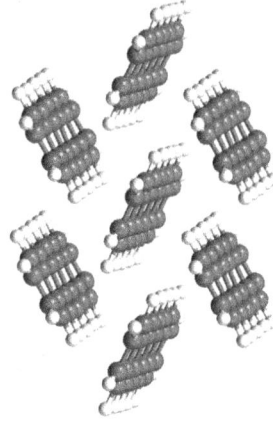

Rubrene

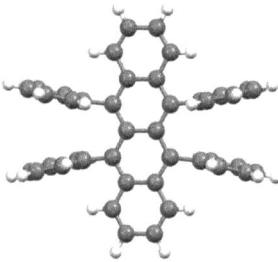
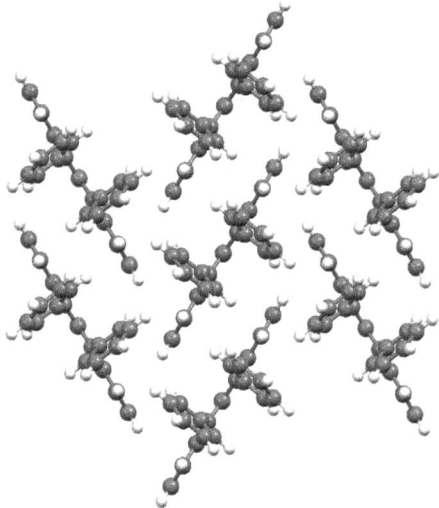

Fig. 6.1   Molecular (left) and crystal (right) structures of pentacene (upper) and rubrene (lower) as representative organic semiconductors.

6.3.   Crystal Structure Calculation of Organic Semiconductors

To elucidate the conduction mechanism of organic semiconductors at the atomistic level, it is necessary to accurately obtain the crystal structure of the materials. First-principles electronic states calculation based on density functional theory is effective as the first step. Inorganic semiconductor materials typified by silicon are bonded with strong covalent bonds between atoms, and these crystal structures can usually be calculated with an accuracy within 1 %. On the other hand, to determine the crystal structure of an organic semiconductor consisting of molecules periodically condensed by weak van der Waals bonds, it is necessary to highly accurately evaluate these weak bonds, and calculation with such high precision is not easy. Recently, first-principles methodologies for evaluating weak bonds by van der Waals forces, such as the van der Waals density functional theory method (vdW-DF)[5] and dispersion-corrected density functional theory (DFT-D),[6] have been developed, and the high-precision crystal structure calculation of organic semiconductors has become feasible in a realistic calculation time. The calculation accuracy has been increased by refining the coefficient of the force inversely proportional to the interatomic distance r to the sixth power or eighth power in the dispersion correction method, and by improving functional such as vdW-DF2[7] and rev-vdW-DF2[8] for use in van der Waals density functional theory. Table 6.1 shows a comparison of lattice constants for rubrene crystals calculated by various methods. By comparison with the results of experiments, it can be seen that it is becoming possible to calculate the crystal structures of organic semiconductors with similar accuracy to that for inorganic materials.

Table 6.1 Calculated and experimental lattice constants (a, b, and c), and binding energy (E_b) for rubrene crystal.[8] The results of density functional calculation with the generalized gradient approximation by Perdew, Burke, and Ernzerhof (PBE), the second version of van der Waals density functional (vdW-DF2), the revised version (rev-vdW-DF2), and experimental values at 100 K are given.

	a (Å)	b (Å)	c (Å)	E_b (eV/mol)
PBE	27.989	7.804	16.585	0.234
vdW-DF2	26.969	7.167	14.523	2.319
rev-vdW-DF2	26.828	7.134	14.04	2.18
Experiment	26.789	7.17	14.211	

6.4.   Transport Theory of Organic Semiconductors

Organic semiconductors exhibit various transport properties depending on the constituent molecules and crystallinity. Figure 6.2 shows the temperature dependences of mobility in some samples. Here, the mobility is an indicator of the conduction, and indicates the degree of carrier transport upon applying a voltage.

As mentioned before, in the low mobility regime in the left figure, thermally activated hopping transport is observed and the mobility exponentially increases with temperature. On the other hand, for recent high-mobility organic semiconductors, the characteristics of band transport are observed as shown in the middle and right figures. The mobility decreases at high temperatures owing to scattering by lattice vibration.

We describe a model of thermally activated hopping transport with low mobility. Carriers are thought to have particle-like properties. They are localized in individual constituent molecules and transfer upon receiving thermal energy from the lattice. Accordingly, the conductivity increases with temperature. In organic semiconductors, when carriers localized in constituent molecules transfer, they move slowly unlike waves, and lattice distortion is considered to be involved in the transfer. Such transfer accompanying the lattice distortion is known as the transfer of quasiparticle polarons with increased effective mass and is described by the hopping model of polaron transport between constituent molecules. A well-known stochastic model for electron

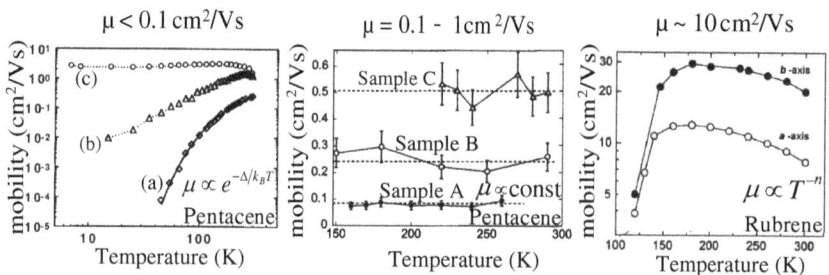

Fig. 6.2   Temperature dependences of mobility μ of organic semicondoctors. (Left: Reprinted from Ref. 9) Thermally activated character $\mu \propto e^{-\Delta/k_B T}$. (Center: Reprinted from Ref. 10) Constant mobility. (Right: Reprinted from Ref. 11) Power $\mu \propto T^{-n}$.

transfer is based on Marcus theory. Using this theory, the carrier transfer probability W_j between molecules is calculated from the intermolecular transfer energy γ_j and reorganization energy λ as

$$W_j = \frac{\gamma_j^2}{\hbar} \sqrt{\frac{\pi}{\lambda k_B T}} \exp\left(-\frac{\lambda}{4 k_B T}\right). \tag{6.1}$$

On the other hand, the diffusion coefficient D is expressed as

$$D \approx \sum_j \Delta R_j^2 W_j P_j, \quad P_j = \frac{W_j}{\sum_k W_k}, \tag{6.2}$$

and the hopping mobility μ is represented by Einstein's relation,

$$\mu = \frac{e}{k_B T} D \propto \exp\left(-\frac{\lambda}{k_B T}\right). \tag{6.3}$$

Next let us consider a conduction model with high mobility. Recent single crystal organic semiconductors have high mobility exceeding that of inorganic amorphous silicon as a result of the elimination of irregularities and grain boundaries. They have a temperature dependence similar to that of the band transport observed in inorganic semiconductors, unlike the thermally activated transport peculiar to hopping transport.[10,11] In band transport, the mobility μ is calculated from the effective mass m^* and relaxation time τ as

$$\mu = \frac{e\tau}{m^*}. \tag{6.4}$$

With increasing temperature, the frequency of scattering increases and the relaxation time of carriers decreases. Since the mobility is proportional to the relaxation time in band transport, the mobility decreases with increasing temperature.

Note that, according to an experiment employing electron spin resonance spectroscopy, a carrier extends spatially over ten molecules.[12] This shows that carrier transport is beyond the scope of Marcus theory, which assume that carriers are localized in the constituent molecules. Moreover, a simple band transport picture is also inadequate, which assumes that carriers extend as waves in a crystal. For the transport analysis of high-performance, high-mobility organic semiconductor materials important for applications, it is necessary to consider various scattering factors such as the generation of polarons and the thermal fluctuation of molecular vibration in addition to the crystal structure. It is expected to analyze the transport properties by a theoretical methodology beyond perturbation theory that regards external factors as additional factors in the scattering process for carrier transport in a soft semiconductor crystal.

We next describe the time-dependent wave packet diffusion (TD-WPD) method[13, 14] as a means of analyzing the transport properties while taking these effects into account in a unified manner. In this method, the motion of carriers is numerically calculated from the time evolution of a wave packet. By forming packets of waves with various wavelengths, it is possible to express particle properties and to analyze the properties of waves as band transport and the properties of particles as hopping transport in a unified manner. We evaluate the mobility μ using the Kubo–Greenwood formula based on linear response theory, which is known as a general theory of electrical conduction, by calculating the time evolution of the coordinates $\hat{z}(t)$ of wave packets of the carriers,

$$\mu = \lim_{t \to +\infty} \frac{q}{n} \int_{-\infty}^{+\infty} dE \left(-\frac{df}{dE} \right) \left\langle \frac{\delta(E - \hat{H}_e)}{\Omega} \frac{\{\hat{z}(t) - \hat{z}(0)\}^2}{t} \right\rangle.$$

(6.5)

Mathematically, the fast calculation of the time evolution operator $\exp(i\hat{H}_e(t)\Delta t/\hbar)$ for Hamiltonian $\hat{H}_e$ on the basis of quantum mechanics is performed using Chebyshev polynomials. In the application to organic semiconductors, we first calculate the effective mass, the intermolecular transfer energy, the reorganization energy, and so forth in accordance with the framework of first-principles calculation based on density functional theory. Next, the time evolution of wave packets of carriers is calculated without assuming a periodic system by combining molecular dynamics simulations describing momently changing molecular motion in a soft organic semiconductor. The relaxation time, mean free path, mobility, and so forth, of carriers are calculated by considering the effects of electron-phonon scattering and polaron generation with lattice distortion accompanying charged carriers, as well as the effect of the thermal fluctuation of molecular vibration. It is also possible to analyze

Polaron

(a)

(0). Input initial conditions at $t=0$

$$|\Psi(0)\rangle \quad \{\Delta R(0)\} \quad \{\Delta u(0)\}$$

(1). Make Hamiltonian of electron at time t

$$\hat{H}_e(t) = \sum_{n,m} \tilde{\gamma}_{nm}^{\text{HOMO}}(\Delta R_{nm}) \, (\hat{c}_n^\dagger \hat{c}_m + \hat{c}_m^\dagger \hat{c}_n) + \sum_n \tilde{\varepsilon}_n^{\text{HOMO}}(\Delta u_n) \, \hat{c}_n^\dagger \hat{c}_n$$

(2). Wave-packet dynamics from t to $t+\Delta t$

$$|\Psi(t)\rangle \rightarrow |\Psi(t+\Delta t)\rangle$$

Evaluation of σ and $D(t)$

- - - - Time-evolution of wavepacket - - - -

$$|\Psi(t+\Delta t)\rangle = \exp\left[i\,\frac{\hat{H}_e(t)}{\hbar}\,\Delta t\right]|\Psi(t)\rangle$$

(3). Molecular dynamics from t to $t+\Delta t$

$$\Delta R(t) \rightarrow \Delta R(t+\Delta t)$$
$$\Delta u(t) \rightarrow \Delta u(t+\Delta t)$$

- - - Equation of motion - - - - - - - - - -

$$M\,\frac{d^2\Delta R}{dt^2} = -\frac{dE_{tot}(\{\rho\},\{\Delta R\},\{\Delta u\})}{d\Delta R}$$

(b)

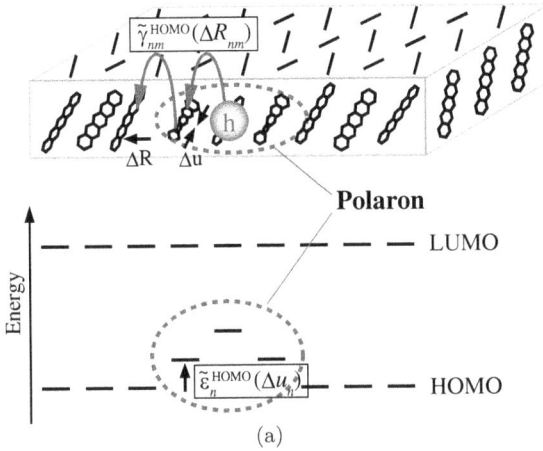

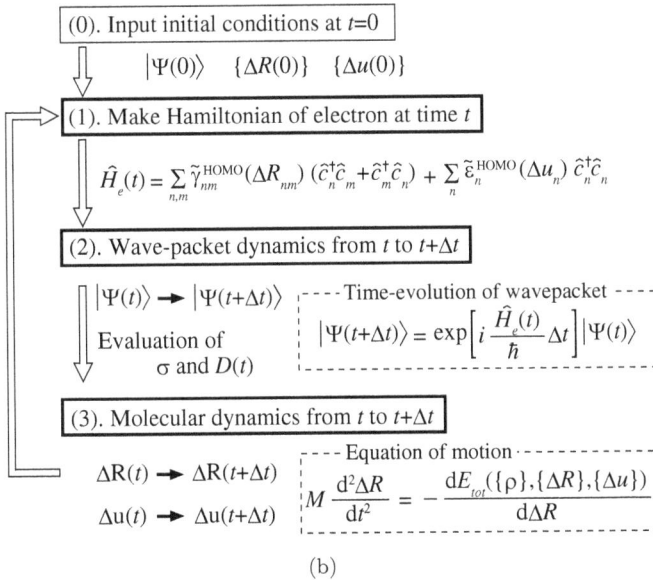

Fig. 6.3 (a) Schematic picture of the carrier transport in organic semiconductors. Carriers are conducted while forming polarons and lattice distorion. (b) Flowchart of the TD-WPD methodology considering dynamical polarons.

the carrier transport mechanism while taking account of various types of scattering such as impurity scattering, as well as the trap potential representing the disorder of the crystal structure (Fig. 6.3). As a result, it is possible to analyze the transition from hopping to band transport

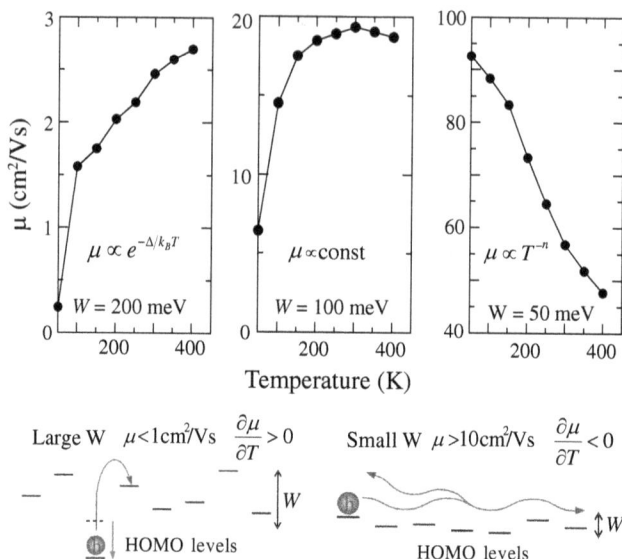

Fig. 6.4 Mobility calculation of organic semiconductor and its temperature dependence by TD-WPD method. The results show hopping transport at low mobilities and bandlike transport characteristics at high mobilities.

and the effect of electron-phonon coupling for polarons described in various types, such as the Holstein and Peierls types, in an organic semiconductor.

Finally, Fig. 6.4 shows the results of a mobility analysis using this TD-WPD method. Here we introduce a single index W to represent the magnitudes of impurities, structure disorder, and, the trap potential, which are present in the experiment, as the disorder of the energy level of the HOMO orbital. When the disorder W is large, the system has a low mobility and the temperature dependence of thermally activated hopping transport. On the other hand, when W is small, the mobility becomes large and has a similar characteristic to band transport. By analyzing the transport properties using the TD-WPD method, it becomes possible to analyze hopping and band transport in a unified manner, which have conventionally been separately investigated. Detailed properties of the carrier transport in organic semiconductors with high mobility can be calculated by considering external factors, such as impurities, in detail.

It is hoped that novel high-performance organic semiconductor devices will be developed using this method.

6.5. Summary

In this chapter, we reviewed the theory of electrical conduction in semiconductor devices, especially the method of calculating electrical conduction based on the first-principles theory and its problems. We also introduced recently developed carrier transport theory for single-crystal organic semiconductors with high mobility. Material development and function control with superior electrical conduction characteristics based on atomistic theory are expected to be further advanced in the future by combining the results of experimental structural analysis along with recent developments in structural calculation methods and transport theory at the atomistic level.

References

[1] M. Tsukada, *et al.*, *J. Phys. Soc. Jpn.* **74**, 1079 (2005).
[2] K. Hirose, H. Ishii, and N. Kobayashi, *J. Vac. Soc. Jpn* **54**, 7 (2011).
[3] K. Hirose and N. Kobayashi, *Quantum Transport Calculations for Nanosystems*, (Pan Stanford, 2014).
[4] Hasegawa and J. Takeya, *Sci. Technol. Adv. Mater.* **10**, 024314 (2009).
[5] M. Dion, *et al.*, *Phys. Rev. Lett.* **92**, 246401 (2004).
[6] S. Grimme, *J. Comput. Chem.* **25**, 1463 (2004).
[7] K. Lee, *et al.*, *Phys. Rev. B* **82**, 081101(R) (2010).
[8] I. Hamada, *Phys. Rev. B* **89**, 121103 (2014).
[9] S. F. Nelson, *et al.*, *Appl. Phys. Lett.* **72**, 1854 (1998).
[10] J. Takeya, *et al.*, *J. Appl. Phys.* **94**, 5800 (2003).
[11] V. Podzorov, *et al.*, *Phys. Rev. Lett.* **93**, 086602 (2004).
[12] K. Marumoto, S. Kuroda, T. Takenobu, and Y. Iwasa, *Phys. Rev. Lett.* **97**, 256603 (2006).
[13] H. Ishii, *et al.*, *Phys. Rev. B* **85**, 245206 (2012).
[14] H. Ishii, N. Kobayashi, and K. Hirose, *Phys. Rev. B* **88**, 205208 (2013).

Chapter 7

INORGANIC MATERIALS

Tomoteru Fukumura
Tohoku University

Daichi Oka
Tohoku University

In this chapter, electronic and magnetic materials are mainly introduced among a large number of inorganic materials. Inorganic materials belong to interdisciplinary fields such as inorganic chemistry,[1] solid state chemistry,[2-5] solid state physics,[6,7] device physics,[8,9] and thin film growth.[10] Structure and functionalities of inorganic materials are closely correlated and are important research topics.

7.1. Inorganic Materials

7.1.1. *Classification of inorganic materials*

(a) Metals, semiconductors, and insulators

Inorganic materials, particularly electronic materials, are classified by their band structures (Fig. 7.1).[6,7] In case of metals, the lower energy band is fully occupied by electrons, and the higher energy band is occupied up to the Fermi level. In case of semiconductors, the valence band at lower energy than the Fermi energy is fully occupied by electrons, whereas the conduction band at higher energy level is unoccupied. There

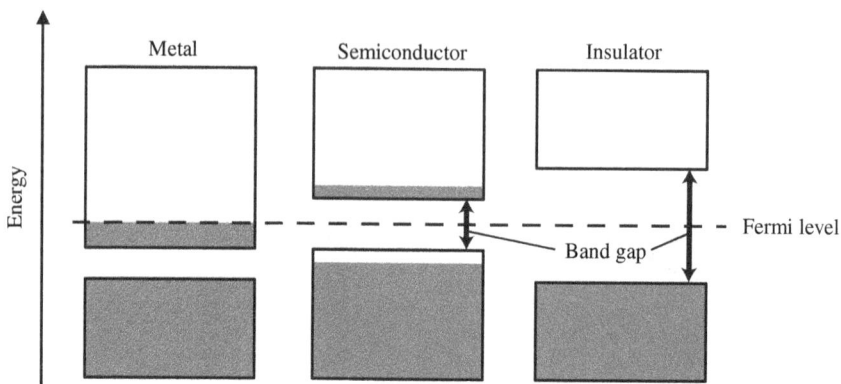

Fig. 7.1   Energy band structure of metal (left), semiconductor (center), and insulator (right).

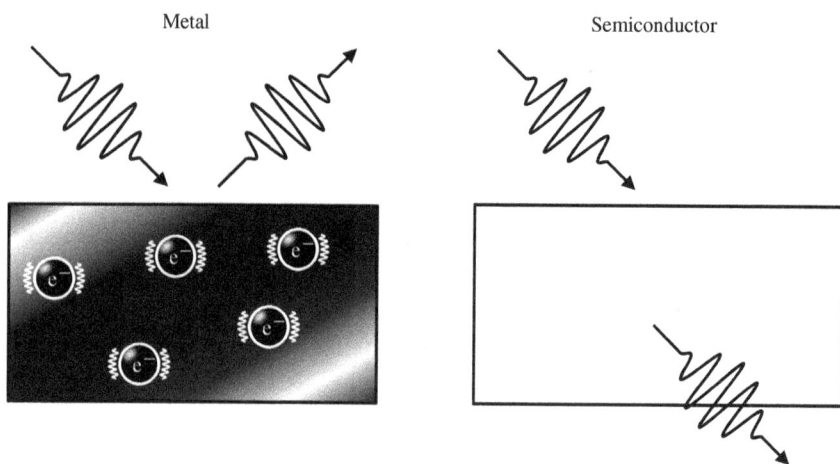

Fig. 7.2   Optical responses of metal (let) and semiconductor (right).

is an energy gap between the conduction and valence bands called as band gap, in which electrons are not allowed to occupy. These band structures govern electrical properties of materials.

High electrical conductivity and metallic luster of metals are caused by free electron carriers at the Fermi level. Under external electric field, free electrons move along the electric field, i.e. electric current flows. Under light irradiation, free electrons are forced to oscillate by ac electric field of the light, resulting in the emission of photon (Fig. 7.2).

On the other hands, electrons of valence band in semiconductor don't move freely under external electric field or light irradiation because of the fully occupied valence band and the presence of band gap between conduction and valence bands. When the photon energy of the light is larger than the band gap, electrons in the valence band are excited to the conduction band. As a result, the excited electrons contribute to electrical conduction. If the photon energy is smaller than the band gap, such excitation will not occur but the light transmits through the material (Fig. 7.2). In case of small band gap, thermal energy will excite electrons in the valance band.

When band gap is larger (>4–5 eV), the optical excitation would be difficult. In such materials called as insulators, electrical conduction is significantly reduced.

(b) Classification by compounds

Inorganic materials can be classified by types of compounds such as simple metals, alloys, borides, carbides, nitrides, oxides, fluorides. Metals and alloys form metallic bonds, and often crystallize in close-packed structures such as cubic and hexagonal ones. Chemical bonds in the other compounds would be covalent or ionic, and construct various crystal structures, depending on the type of compounds. Compounds with similar chemical formula tend to possess a similar crystal structure. For example, there are various ABO_3 (A: alkali metal or alkali earth metal; B: transition metal or main group metal) oxides with a large number of combinations of A and B, and most of them form perovskite structure (Fig. 7.3). Since crystal structure is connected with the band structure, the compounds with the same crystal structure are likely to show similar physical properties.[5]

7.1.2. *Properties of inorganic materials*

(a) Electrical properties

Electrical conductivity σ is an indicator of high electrical conduction, expressed as,[6,7]

$$\sigma = ne\mu,$$

in which, n is the number of conduction electrons per unit volume, e is elementary charge and μ is the electron mobility. Electron mobility

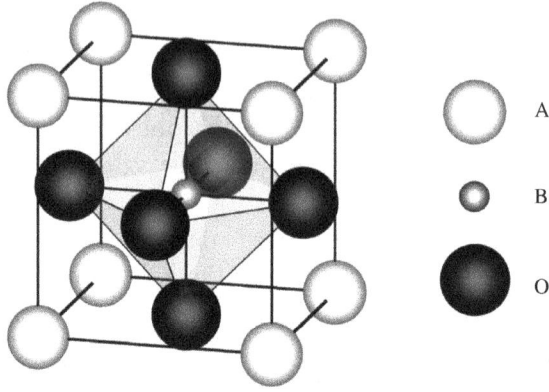

Fig. 7.3   Crystal structure of perovskite type metal oxide ABO_3.

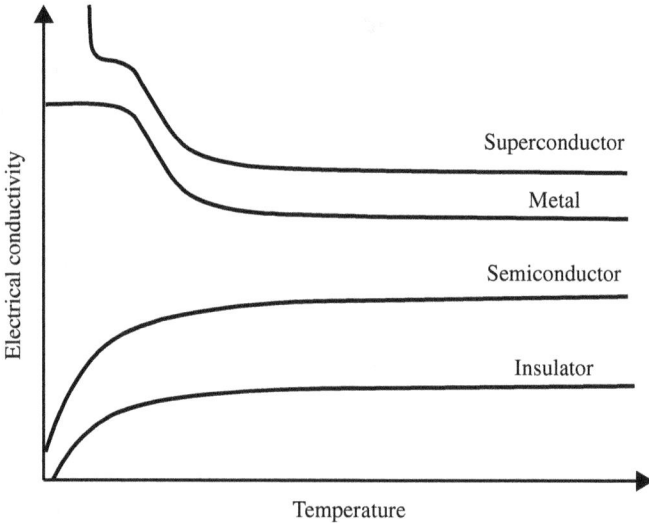

Fig. 7.4   Temperature dependence of electrical resistivity for metal, semiconductor, insulator, and superconductor. The axes are not to scale.

is defined as velocity of electron per electric field, representing the movability of conduction electrons. The electrical resistivity is the inverse of the electrical conductivity.

For metals, electron scattering by lattice vibration is reduced with decreasing temperature. Accordingly, electrical conductivity monotonically increases with decreasing temperature (Fig. 7.4). For semiconductors, the number of thermally excited electron carriers is much smaller than that of conduction electrons in metals. In addition, less electrons are excited at lower temperature, thus the electrical conductivity decreases with decreasing temperature. For insulators, negligible number of electron carriers yields in much smaller electrical conductivity (Fig. 7.4).

Metals have high and invariable electrical conductivity because n and μ are material constants, in principle. On the other hand, semiconductors have variable electrical conductivity because n is variable through electric field effect and chemical doping. This nature is very important for semiconductor devices used in modern electronics.

(b) Superconductivity

Electrical resistance in metals is very low but usually nonzero at low temperature. However, some metals show zero resistance at low temperature (Fig. 7.4). This state corresponds to superconducting phase, and such metals are called as superconductors. In normal conducting state, conduction electrons flow randomly in metals. In superconducting state, electron pair called as the Cooper pair flows coherently in superconductors (Fig. 7.5). The electron is a fermion subject to the Pauli

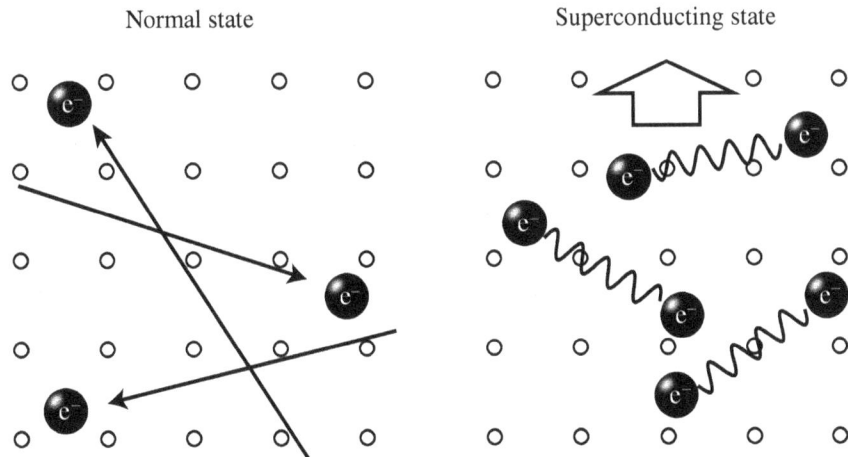

Fig. 7.5   Electrical conductions of normal state (left) and superconducting state (right) in metal.

exclusion principle. On the other hand, the Cooper pair is in a condensed states due to their bosonic nature. As a result, superconductors show the perfect diamagnetism called as the Meissner effect, in which the magnetic flux is expelled from superconductors, and the macroscopic quantum interference phenomena such as the Josephson effect.[11]

Superconductivity is an interesting phenomenon from the viewpoint of basic science, but also is important from industrial viewpoint, because electrical current without power consumption in zero resistance state can be used for electric power storage and high magnetic field production. One big problem is that superconductivity is available only at low temperature. Today, researchers are trying to find high temperature superconductors.

(c) Magnetism

Magnetic properties of inorganic materials are also important. Spins of electrons are the origin of magnetism, and various magnetism is generated by exchange interaction between spins. Spin moment is canceled out in compounds without unpaired electrons, so that the resultant magnetism is diamagnetism. For example, Al_2O_3 shows diamagnetism since Al^{3+} and O^{2-} ions in Al_2O_3 possess the closed shell. On the other hand, compounds with unpaired electrons have nonzero spin magnetic moments, thus show paramagnetism. Exchange interaction between the spins, if any, forces the spins parallel or antiparallel. The former and the latter correspond to ferromagnetism and antiferromagnetism, respectively. A derivative of the antiferromagnetism is ferrimagnetism, in which the antiparallel spins have different magnitude. Each magnetism is schematically drawn (Fig. 7.6). Special feature of ferromagnetism and ferrimagnetism is the large magnetization — magnetic moments per unit volume — even without external magnetic field. Accordingly, ferromagnetic and ferrimagnetic materials are used for various applications such as permanent magnets, motors, magnetic recording media, and optical isolators.[12]

Interplay between spin and charge, which are degrees of freedom of electron, produces giant magnetoresistance in magnetic multilayers and tunnel magnetoresistance in magnetic tunneling junctions, whose electrical resistance significantly depends on magnetization directions in

Paramagnetism   Ferromagnetism

Antiferromagnetism   Ferrimagnetism

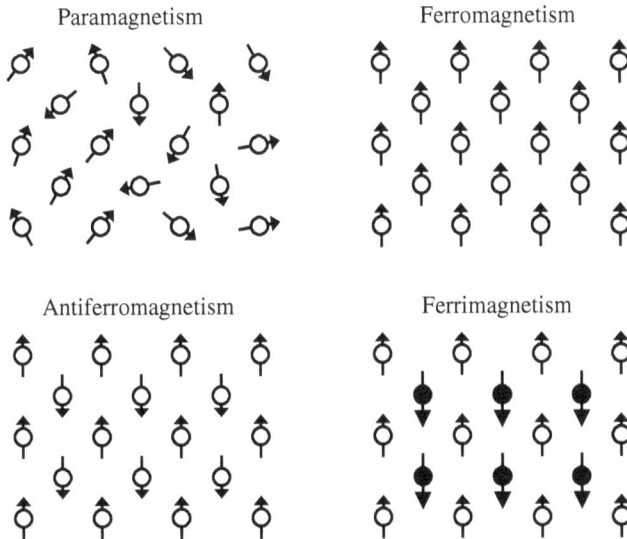

Fig. 7.6   Schematic spin arrangements of paramagnetism, ferromagnetism, antiferro-magnetism, and ferrimagnetism.

the magnetic layers. These emerging research field is called as spintronics, and novel magnetic memories and sensors are developing.[13]

(d) Various inorganic functional materials

As described above, a large number of inorganic materials are developed even limited to electrical and magnetic properties. Table 7.1 shows an example of applications of inorganic materials.

7.2.   Synthetic Methods of Inorganic Materials

In this section, representative synthetic methods for various forms of inorganic materials, mainly focusing on oxides, are described.

7.2.1.   *Bulk materials*

Bulk material denotes its size larger than several millimeters. Bulk synthesis has been performed for a long time. Oxides are often synthesized in air. For example, perovskite-type $BaTiO_3$ is synthesized in air by

Table 7.1   Structures, properties and applications of inorganic materials.

Type	Compound	Composition	Structures at r.t.[*1]	Properties	Applications
Metals	Aluminum	Al	Face-centered cubic	Electrically conductive	Conductor
	Iron	Fe	Body-centered cubic	Electrically conductive, ferromagnetic ($T_C = 1043$ K)	Structural material
	Copper	Cu	Face-centered cubic	Electrically conductive	Conductor
	Silver	Ag	Face-centered cubic	Electrically conductive	Electrode, mirror
	Platinum	Pt	Face-centered cubic	Electrically conductive, corrosion resistant	Electrode, catalyst, crucible
	Gold	Au	Face-centered cubic	Electrically conductive, ductile	Electrode, catalyst
Alloys[*2]	Brass	$Cu_{0.7}Zn_{0.3}$	Cu	Electrically conductive, ductile	Structural material
	Nichrome	$CrNi_3$	Cu_3Au	High electrical resistivity ($\rho = 1.28 \times 10^{-6}$ Ωm)	Heating wire
	Permalloy	$FeNi_3$	Cu_3Au	Electrically conductive, ferromagnetic ($T_C = {\sim}1180$ K)	High permeability for sensor and magnetic shielding
Borides	Magnesium diboride	MgB_2	AlB_2	Superconducting ($T_c = 39$ K)	Superconducting coil
	Titanium diboride	TiB_2	AlB_2	Electrically conductive, high hardness	Protection material

Table 7.1   (*Continued*)

Type	Compound	Composition	Structures at r.t.*[1]	Properties	Applications
	Lanthanum hexaboride	LaB_6	CaB_6	Electrically conductive, low work function	Hot cathode
	Rhenium diboride	ReB_2	ReB_2	Electrically conductive, high hardness	Superhard material
Carbides	Vanadium carbide	VC	Rock-salt	Electrically conductive, high hardness	Superhard material
	Niobium carbide	NbC	Rock-salt	Electrically conductive, high hardness	Superhard material
	Tungsten carbide	WC	WC	Electrically conductive, high hardness	Superhard material
Nitrides	Lithium nitride	Li_3N	Li_3N	Lithium ion conducting	Hydrogen storage
	Gallium nitride	GaN	Wurtzite	Band gap ~3.4 eV	Semiconductor
	Titanium nitride	TiN	Rock-salt	Electrically conductive, high hardness	Protection material
	Manganese zinc nitride	Mn_3ZnN	Antiperovskite	Antiferromagnetic ($T_N \sim 170$ K)	Negative thermal expansion material
	Niobium nitride	NbN	Rock-salt	Superconducting ($T_c = 16$ K)	High speed infrared receiver
Oxides	Magnesium oxide	MgO	Rock-salt	Insulating	Insulator in magnetic tunneling junction
	Titanium dioxide	TiO_2	Rutile	Band gap ~3.0 eV	Photocatalyst

Table 7.1 (*Continued*)

Type	Compound	Composition	Structures at r.t.[*1]	Properties	Applications
	Strontium titanate	$SrTiO_3$	Perovskite	Band gap ~3.2 eV, quantum paraelectric	Dielectric, photocatalyst, substrate
	Indium tin oxide	$In_{2-x}Sn_xO_3$	Bixbyite	Electrically conductive	Transparent electrode
	Yttrium barium cuprate	$YBa_2Cu_3O_{7-\delta}$	Oxygen deficient perovskite	Superconducting (T_c = 94 K)	Superconducting quantum interference devices
	Rhenium oxide	ReO_3	ReO_3	Electrically conductive	Catalyst
Fluorides	Fluorite	CaF_2	Fluorite	Insulating, band gap ~9.4 eV	Optical material
	Cryolite	Na_3AlF_6	$[NH_4]_3FeF_6$	Glassy luster	Refining flux of aluminum metal
Others	Silicon	Si	Diamond	Band gap ~1.2 eV	Semiconductor
	Germanium	Ge	Diamond	Band gap ~0.7 eV	Semiconductor
	Gallium arsenide	GaAs	Zincblende	Band gap ~1.4 eV	Semiconductor

[*1]r.t. stands for room temperature. [*2]Typical compositions and structures are shown.
T_C: Curie temperature. T_C: superconducting critical temperature.

sintering the following powder agents in a furnace.

$$BaCO_3 + TiO_2 \rightarrow BaTiO_3 + CO_2(gas)$$

Because oxides have usually high melting points, their synthesis requires relatively high sintering temperature. The above reaction is a solid phase reaction generally producing a polycrystal. Czochralski method is a single crystal growth method by pulling up melted specimen. Hydrothermal method is also a single crystal growth method using high temperature and pressure in autoclave with water and reactants for synthesis of e.g. quartz and zinc oxide. Single crystals are highly crystalline free from grain boundaries and defects, thus show intrinsic properties not influenced by crystalline structure. Highly pure single crystals are obtained with purification process, and are important factor for modern electronics.

Transition metal elements have several oxidation states, some of which could be unstable at high temperature in air. In such cases, synthesis is performed in a reductive atmosphere, e.g. in evacuated quartz tube. The Ellingham diagram indicates stable oxidation state under a specific temperature and partial pressure of various gasses (oxygen, hydrogen, etc.), being useful for deciding synthesis condition.

7.2.2. *Thin films*

Inorganic materials in electronic devices are required to have a definite thickness, and the smaller size is preferable for device integration. In this case, not bulk materials but thin film materials with nanometers or micrometers thick are used. The lateral size is controlled by microfabrication techniques such as photolithography and electron beam lithography. Also, it is possible to fabricate multilayers and heterostructures in addition to the heavy impurity doping beyond thermodynamic solubility limit and the stabilization of metastable crystal structure.

Sol-gel method is a film growth technique to sinter reactants dispersed in colloidal gel casted on substrate under atmospheric pressure. Although this technique is facile, resultant thin film is usually low crystalline polycrystal. On the other hand, thin films deposited in vacuum chamber are free from many impurities. When the vacuum level is high, chemical species vaporized from the starting materials reach on the substrate as a molecular beam without any collision with residual gas molecules. Use of lattice matched single crystal substrates enables to obtain epitaxial thin films at much lower growth temperatures than those for bulk crystals. Molecular beam epitaxy (MBE) is often used for thin film deposition of compound semiconductors such as GaAs. Here, highly crystalline target compounds can be obtained on substrates by individually supplying each constituent element. Pulsed laser deposition (PLD) is a thin film growth technique using laser ablation of a starting material with the same composition as the target compound. With PLD, compounds with high melting point can be easily deposited by using focused ultraviolet laser, and highly crystalline epitaxial thin films can be obtained on lattice matched substrates. In case of oxides, oxygen stoichiometry is controlled by supplying oxygen or ozone gas during film deposition. In chemical vapor deposition (CVD), thin films are deposited by supplying volatile reactants onto the substrates in a reaction tube (Fig. 7.7).

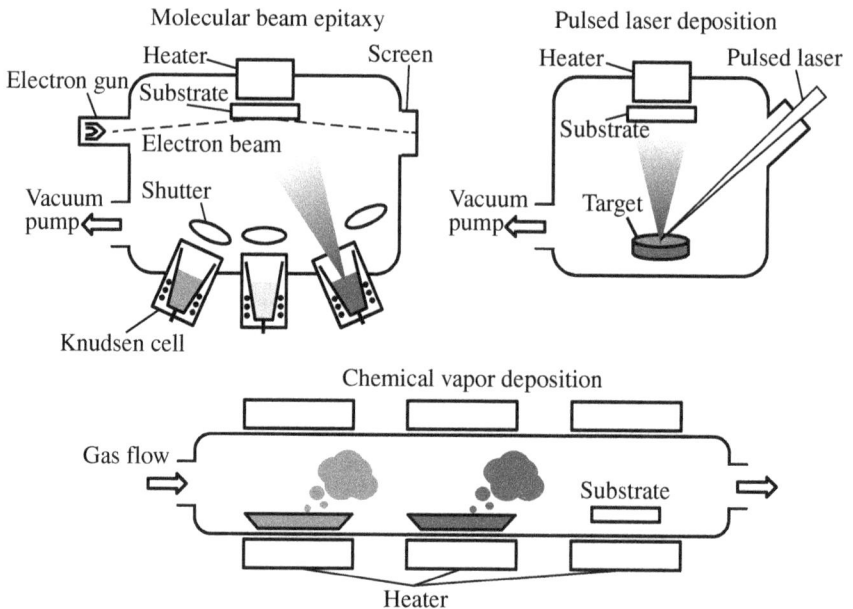

Fig. 7.7 Schematic figures of molecular beam epitaxy, pulsed laser deposition, and chemical vapor deposition.

7.3.   Structural Analyses of Inorganic Materials

7.3.1.   *Crystal structure*

The syntheses of inorganic materials described in Section 7.2 are usually followed by their structural characterizations. Diffraction methods are useful to identify periodic crystal structures. The most standard among them is X-ray diffraction (XRD).[14] By measuring XRD pattern, information of crystal structure, crystal orientation, crystallinity, and presence of impurity phase can be obtained. In case of epitaxial thin film on single crystal substrate, crystallographic relation between the thin film and substrate can be examined. However, XRD is not suitable for minute samples such as nanometer thick ultrathin films and amorphous samples lacking periodic lattice structure.

Crystal structure can be identified by neutron diffraction, too. Because neutron beam is strongly scattered by light elements, being different from X-ray, light atoms such as hydrogen can be probed. Also, magnetic structure, that is spin arrangement, can be measured via neutron's spins. However, large sample volume is necessary for neutron measurements because of the weak interaction between neutron beam and materials.

7.3.2.   *Surface structure*

Although X-ray and neutron diffraction methods are useful to evaluate crystal structures of bulk specimens, surface sensitive measurements with high spatial resolution are also available for characterization of surface structures. Various analytical methods have been developed owing to modern nanotechnology. Prevailing optical microscope possesses a spatial resolution down to submicrometers.

Scanning electron microscope (SEM) uses electron beam with much shorter wavelength than visible light, thus has a spatial resolution of nanometers with wide dynamic range. By attaching energy- or wavelength-dispersive X-ray spectrometer, composition analysis and elemental mapping can be performed from X-ray fluorescence spectra of samples generated by electron beam irradiation. However, SEM is not suitable for the measurement on insulators because of charge-up effect.

Atomic force microscope (AFM) maps atomic forces between cantilever and sample surface. Lateral and vertical spatial resolutions are an order of nanometers and subnanometers, respectively, suitable for observation of nanostructures. Scanning tunneling microscope (STM) measures tunneling current between metal tip and sample surface, enabling to observe atoms directly, in addition to density of states in an atomic scale.

Reflection high energy electron diffraction (RHEED) is used to observe topmost surface atomic structure. This method is surface sensitive because of strong elastic and inelastic scattering of electron beam at surface. Surface flatness and structure are evaluated from the diffraction pattern by irradiating high speed electron beam with shallow incident angle. This method is often used for in-situ observation on surface structure and growth mode during film deposition.

7.3.3. Local structure

Transmission electron microscope (TEM) is used to observe local structure down to atomic scale, in which the sample needs to be processed in a form of sufficiently thin flake for transmission of electron beam. In addition, electron beam diffraction, energy-dispersive X-ray spectroscopy, and electron beam energy loss spectroscopy (EELS) can be performed for microscopic evaluation of local structure, local composition, and elemental identification on an atomic site. Scanning transmission electron microscope (STEM) enables to observe element-dependent atomic image because of the higher contrast for heavier elements (Fig. 7.8).

Extended X-ray absorption fine structure (EXAFS) is one of the X-ray absorption spectroscopy techniques, and evaluates coordination number and bond length around specific atomic site in the sample. X-ray fluorescence holography (XFH) can construct three dimensional

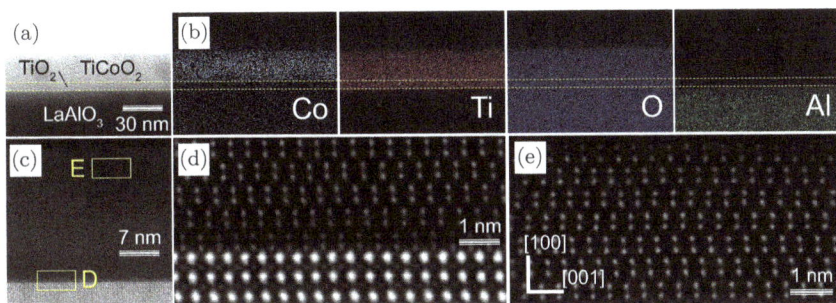

Fig. 7.8 (a) Bright field STEM image of anatase Co doped TiO_2 thin film/TiO_2 buffer layer/$LaAlO_3$ substrate. (b) Energy-dispersive x-ray spectroscopy mappings for each element at region A. (c) Cross sectional STEM image of thin film sample. (d and e) High resolution HAADF-STEM images at regions D and E (HAADF: High-angle annular dark-field).[19]

atomic structure around specific atomic site by extending the principle of EXAFS, and is, for example, used to evaluate local substructure surrounding a small amount of dopant atoms as described below.

7.4. Relation between Structure and Functionalities: Transition Metal Doping into Oxide Semiconductor

Chemical doping is an important technique to generate various functionalities in semiconductors. Electron and hole carriers are generated by doping donor and accepter elements, respectively. Formation of solid solution from compound group with the same crystal structure such as AlAs-GaAs-InAs realizes tunable band gap and lattice constants as a function of the composition. Here, we show transition metal doping into oxide semiconductor.

Not only conduction carriers but also localized spins are introduced in semiconductors by transition metal doping. Even in case of the small amount of doping, the carriers intermediate exchange interaction between spatially separated localized spins of the transition metal, because the carriers (electrons or holes) also possess their spins. Such indirect exchange interaction between the localized spins may generate ferromagnetism. For example, Mn-doped GaAs shows ferromagnetism with the Curie temperature of about 200 K. Such coexistence of ferromagnetic and semiconducting properties is suitable for semiconductor spintronics, since electric field control of ferromagnetism is possible.[15]

Apparent from the fact that various oxide semiconductors are used for transparent electrodes, the oxide semiconductors generally possess both wide band gap and high electron density. Heavy electron mass in such wide gap semiconductors is expected to generate strong exchange interaction between localized spins, resulting in a high Curie temperature by doping transition metal. After material exploration based on such strategy, Co-doped TiO_2 was found to be room temperature ferromagnetic semiconductor with Curie temperature of 600 K at maximum.[16] There is no satisfactory theoretical explanation for the high temperature ferromagnetism, however, the intrinsic ferromagnetic properties have been experimentally confirmed such as the appearance of anomalous Hall effect and large magneto-optical effect[17,18] and the electric field induced room temperature ferromagnetism.[19,20] Several research groups attributed the high Curie temperature to defect-mediated exchange interaction because of the lack of suitable mechanism.[21]

Three dimensional local atomic structure around Co in rutile Co-doped TiO_2 was measured by XFH. Conventional scenario was that Ti is just replaced by Co preserving the rutile structure, however, the actual local atomic structure was different from the rutile structure in the ferromagnetic Co-doped TiO_2 (Fig. 7.9). While the paramagnetic Co-doped TiO_2 with a low Co content possessed usual rutile structure (Figs. 7.9(a) and (c)), the ferromagnetic Co-doped TiO_2 with a high Co content possessed a suboxidic structure with smaller oxygen coordination numbers (Figs. 7.9(b) and (d)). Because it is difficult to identify a large scale structure and light elements such as oxygen with XFH, the first-principles calculation is helpful to identify the total crystal structure. From the

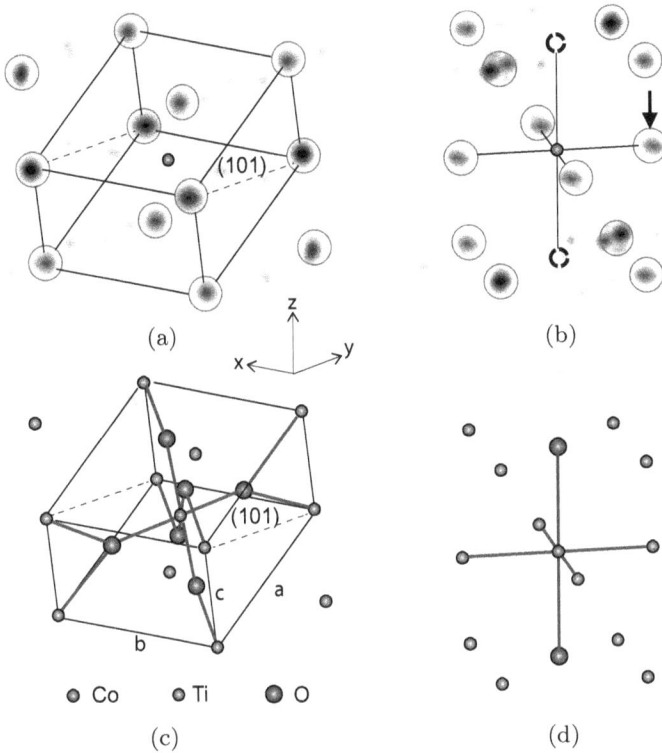

Fig. 7.9 Three dimensional local structure around Co atom in rutile Co-doped TiO_2 measured by x-ray fluorescence holography[22] for (a) paramagnetic $Ti_{0.99}Co_{0.01}O_2$ and (b) ferromagnetic $Ti_{0.95}Co_{0.05}O_2$ thin films. (c) and (d) Structural models deduced from (a) and (b). (Reprinted from Applied Physics Letters **106**, 222403 (2015), with the permission of AIP Publishing.)

calculation, the seemingly unstable suboxidic structure was found to be embedded in rutile structure (Fig. 7.10). Such suboxidic structure was not detected by XRD and TEM because of the small volume, representing the advantage of XFH to measure three dimensional local atomic structure around dopants.

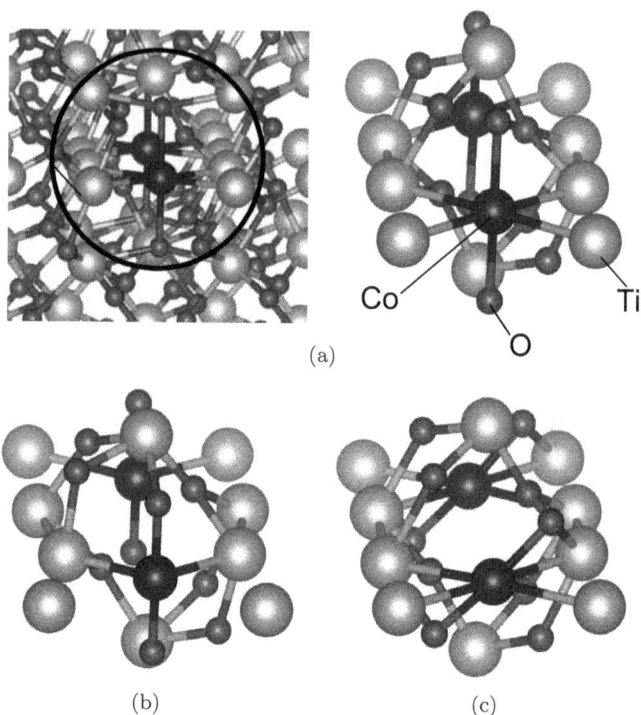

(a)

Co   Ti

O

(b)   (c)

Fig. 7.10 Suboxidic structure around Co atom in rutile Co-doped TiO_2 estimated from x-ray fluorescence holography measurements and first-principles calculation.[22] (Reprinted from Applied Physics Letters **106**, 222403 (2015), with the permission of AIP Publishing.)

Acknowledgement

Crystal structure in Fig. 7.3 was drawn by the VESTA program.[23]

References

[1] M. Weller, *et al.*, *Inorganic Chemistry* Oxford University Press (2014).
[2] L. Smart and E. Moore, *Solid State Chemistry: An Introduction* (CRC Press, 2017).
[3] A. R. West, *Basic Solid State Chemistry* (Wiley, 1999).
[4] P. A. Cox, *The Electronic Structure and Chemistry of Solids* (Oxford Science Publications, 1987).
[5] F. S. Galasso, *Structure and Properties of Inorganic Solids* (Pergamon, 1970).
[6] C. Kittel, *Introduction to Solid State Physics* (Wiley, 2005).

[7] N. W. Ashcroft and N. D. Mermin, *Solid State Physics* (Cengage learning Asia, 2016).

[8] A. S. Grove, *Physics and Technology of Semiconductor Devices* (Wiley, 1967).

[9] S. M. Sze and K. K. Ng, *Physics of Semiconductor Devices* (Wiley-Interscience, 2006).

[10] D. L. Smith, *Thin-Film Deposition: Principles and Practice* (McGraw-Hill Education, 1995).

[11] M. Tinkham, *Introduction to Superconductivity* (Dover Publications, 2004).

[12] S. Chikazumi, *Physics of Ferromagnetism* (Oxford University Press, 2009).

[13] S. Bandyopadhyay and M. Cahay, *Introduction to Spintronics* (CRC Press, 2015).

[14] B. D. Cullity, *Elements of X-Ray Diffraction* (Addison-Wesley, 1978).

[15] H. Ohno, *Science* **281**, 951 (1998).

[16] Y. Matsumoto, *et al.*, *Science* **291**, 854 (2001).

[17] H. Toyosaki, *et al.*, *Nat. Mater.* **3**, 221 (2004).

[18] T. Fukumura, *et al.*, *Jpn. J. Appl. Phys.* **42**, L105 (2003).

[19] Y. Yamada, *et al.*, *Science* **332**, 1065 (2011).

[20] T. Fukumura and M. Kawasaki, "Functional Metal Oxides: New Science and Novel Applications" ed. by S. Ogale, M. Blamire, T. Venkatesan, Wiley, (2013) p91.

[21] K. Griffin Roberts, *et al.*, *Phys. Rev. B* **78**, 014409 (2008).

[22] W. Hu, *et al.*, *Appl. Phys. Lett.* **106**, 222403 (2015).

[23] K. Momma and F. Izumi, *J. Appl. Crystallogr.* **44**, 1272 (2011).

Chapter 8

MATERIALS FOR ORGANIC ELECTRONICS AND BIOELECTRONICS

Subchapter 8.1

MATERIALS FOR ORGANIC ELECTRONICS AND BIOELECTRONICS

Yoshihiro Kubozono
Okayama University

Hidenori Goto
Okayama University

Hiroko Yamada
Nara Institute of Science and Technology

In this chapter, the principles behind the mechanism of operation of organic transistors are described, as they are both important in their own right and also a representative example of organic electronics. Accordingly, the recent progress in organic transistors will be explored in depth. The term "organic transistor" generally refers to the field-effect transistor (FET), which uses thin films of organic materials. The study of these "organic FETs" began in the 1980s.[1] Initially, the available field-effect mobility (μ), which is one determinant of the performance of an FET, was ca. 10^{-5} cm^2 V^{-1} s^{-1}, a value that is insufficient for practical applications. In the late 1990s, it was found that pentacene, in which five benzene rings are connected in a linear structure (Fig. 8.1.1), and C$_{60}$, in

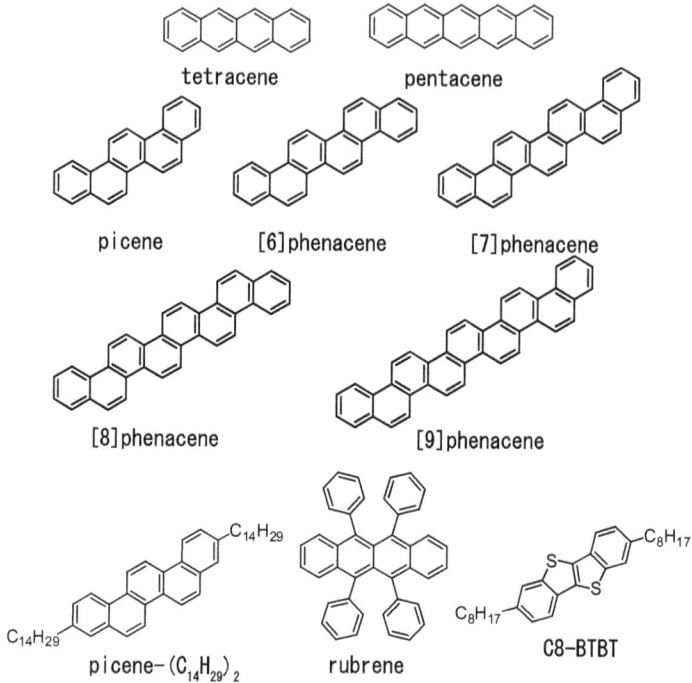

Fig. 8.1.1 Molecular structure of representative organic materials used for active layer of organic FET.

which 60 carbon atoms form a spherical structure, were good candidates for the active layer of an FET device. The μ value became ~1 cm^2 V^{-1} s^{-1} in FETs using these molecules.[2,3] This μ value raised the expectation that organic FETs might in the future become practical devices.

On the other hand, since the year 2000, some work on FETs has demonstrated that the physical properties of materials can be controlled by carrier accumulation, such as the field-induced carrier doping of C$_{60}$ for inducing superconductivity. These studies were not the case, but the interest in field-induced carrier doping to control the physical properties of materials has attracted much attention from researchers in solid-state physics. Consequently, some control of the physical properties of inorganic materials has been achieved by field-induced carrier accumulation using a device with the structure of an FET.[4,5]

Research on organic FETs has been brought to a new stage by the participation of many researchers from other fields, where FET performance was pursued not only to find practical applications, but also to investigate fundamental physics. In 2003, a device with excellent FET properties that used a single crystal was fabricated using rubrene (Fig. 8.1.1).[6] This showed that transistor operation in an organic FET could be achieved using not only thin films but also single crystals. An FET using a single crystal is superior to a thin-film FET for studying the intrinsic nature of organic materials because many interfering factors such as structural defects, grain boundaries, and impurities can be removed from the active layer. Generally, FET performance is also higher in a single-crystal FET than in a thin-film FET. Clearly, these studies have led to remarkable advances in the research on organic FETs. Furthermore, Hall-effect measurements were achieved in single crystals of organic molecules,[7,8] and angle-resolved photoemission spectroscopy (ARPES) was performed on some organic single crystals.[9,10] Based on these results, it was suggested that electron transport in organic single-crystal FETs occurs by band transport rather than hopping transport. Since then, the idea that electron transport in an organic thin-film FET is explained by simple hopping transport has been modified due to analyses based on the multiple shallow trap-and-release (MTR) model for the temperature-dependence of transport properties in organic thin-film FETs exhibiting a high μ value.[11-13] The MTR model assumes that the carriers in a band are captured by trap states and thermally excited up to the conduction (or valence) band, i.e. this concept is substantially based on band transport.

Furthermore, the reduction of μ value caused by thermal scattering of phonons was determined from the temperature-dependence of μ in organic single-crystal FETs.[14,15] This constituted the direct observation of band transport in organic single-crystal FETs. These studies have enabled the full discussion of the mechanism of organic FET operation. An FET using pentacene (Fig. 8.1.1) as the active layer shows typical p-channel operation with Au source/drain electrodes, but shows n-channel operation with Ca source/drain electrodes.[16] So an organic FET can operate both as p-channel and n-channel depending on the electrode metal, a flexibility termed "ambipolar". Thus, an organic FET can operate in an ambipolar way by changing the metal of the source/drain electrodes (or changing the work function of the electrodes), unlike the operation of Si MOSFETs, which can operate in either p-channel or n-channel through the formation of an inversion layer in which minority carriers in the bulk become the majority in the channel region. Also, neither organic thin films nor crystals are doped with any impurities, unlike the Si crystal used in Si MOSFETs, which can be doped with an acceptor (p-type Si) or with a donor (n-type Si) material. In Section 8.1.1, we will fully examine organic FETs based on the background described above.

8.1.1. Evaluation of the Characteristics of Organic FETs

Initially, FET operation was explained based on the general theory of MOSFETs. Si has four covalent bonds, and displays a diamond-type crystal structure. The local structure around Si is tetrahedral. Such arrangements are observed in C and germanium (Ge), as in the diamond structure of C. The bond between Si atoms is a covalent bond, which is stronger than other types of bonds. Because of the strong bonds, the Si crystal can produce a band based on the orbitals of the Si atom. Therefore, the understanding of FET properties must be based on energy band theory. Pure Si has no impurities, and Si crystals doped with impurities are used as the active layers of FET devices. Non-doped Si is an 'intrinsic semiconductor', which has a 1.12 eV band gap in air. The charge carriers responsible for electrical currents in metals are simply electrons (or, strictly speaking, holes for Al and In). On the other hand, both electrons and holes can become carriers in a semiconductor. However, few carriers (electrons or holes) are present in an intrinsic semiconductor, where they are produced by thermal excitation. The concentration of electrons and holes is the same in an intrinsic semiconductor. When applying a bias voltage, the electrons and holes move in opposite directions, which produces an electric current flowing in the same direction. Therefore, both carriers (electrons and holes) play a significant role in electric transport in a semiconductor; the concentration of electrons is much higher in metals, and the presence of holes is negligible.

In intrinsic semiconductors, as seen from Fig. 8.1.2, the number of electrons (n) and holes (p) is the same.

$$n = p. \tag{8.1.1}$$

conduction band

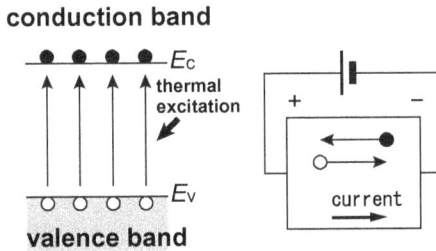

Fig. 8.1.2   Carrier generation and direction of electric current flow. Solid and open circles represent electrons and holes, respectively.

We define the intrinsic carrier concentration (n_i) as follows:

$$np = n_i^2. \tag{8.1.2}$$

and n_i will be referred in the discussion of extrinsic semiconductors. Equation (8.1.2) holds for any semiconductor at thermal equilibrium, and is called the "law of mass action" as applied to electronics. The Fermi energy, E_F, of an intrinsic semiconductor would appear to be expressed by the following equation, because $n = p$:

$$E_F = \frac{E_c + E_v}{2}, \tag{8.1.3}$$

where E_c and E_v refer to the lowest-energy of conduction band and the highest-energy of valence band, respectively. However, the actual E_F differs from Eq. (8.1.3), and is given exactly by:

$$E_F = \frac{E_c + E_v}{2} + \frac{3}{4}k_B T \ln \frac{m_h^*}{m_e^*} \equiv E_i. \tag{8.1.4}$$

The effective electron mass, m_e^*, is generally different from the effective hole mass, m_h^*, in the crystal, so E_F is not centred between E_c and E_v. Here, the subscripts e and h refer to electron and hole, and '*' implies that the mass is not static mass, but mass in the crystal. The term k_B is Boltzmann's constant. As seen from Eq. (8.1.4), the E_F is expressed by Eq. (8.1.3), when $T = 0$ or $m_e^* = m_h^*$. The E_i in Eq. (8.1.4) is the Fermi energy of an intrinsic semiconductor, which is called its "intrinsic energy".

The semiconductor, when doped with impurities, is called an "extrinsic semiconductor". Dopant atoms (or impurities) are present in quantities on the order of 1 per 10^6 Si atoms, so dopant atoms are completely isolated from other dopant atoms. The atoms used for impurity doping belong to group 13 or 15 in the periodic table, i.e. the groups adjacent to Si (group 14). If Si is doped with a B atom (from group 13), a hole is doped to Si because of the structural lack of one electron, while if Si is doped with P or As atoms (from group 15), an electron is doped to Si, because of the structural excess of one electron. The doped (or injected) electrons and holes can freely move to constitute a flow of electric current when subjected to a bias voltage. Figure 8.1.3 shows the energy diagram of an electron-doped semiconductor. This energy diagram corresponds to an extrinsic semiconductor in which Si was doped with P or As atoms, with its excess electrons.

For example, the number of electrons in an As atom is more by one than in Si, and the electrons excited from the donors, i.e. the atoms that can donate electrons, occupy the conduction band (see Fig. 8.1.3). A donor such as As becomes As^+. Here it should be noted that the energy levels of donors are denoted by discrete lines, because the donors cannot form a band owing to their isolation from other donors. Many electrons are excited from the As atoms to the conduction band because the energy difference between the energy level of As and E_c is much smaller than that between E_v and E_c. Therefore, $n \gg p$ in this case, and As becomes As^+ after an electron is excited to the conduction band. Here, it should be noted that the As^+ cannot move through the structure at room temperature because of its significant mass. Consequently, As^+

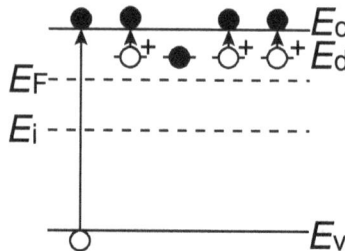

Fig. 8.1.3   Carrier generation in an n-type semiconductor.

plays no role in any electric current, though an electric current may be formed by ions at high temperatures, a phenomenon known as 'ionic conduction'. Thus, when Si is doped with As, the following chemical formula holds,

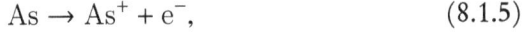

$$As \rightarrow As^+ + e^-, \quad (8.1.5)$$

and electrons are donated to Si. Such a semiconductor, with electrons as majority carriers, is called an "n-type semiconductor". If considering that the E_F is defined as the energy that the Fermi-Dirac function, $f(E)$, expressing the probability of electron distribution is 1/2, the E_F shown in Fig. 8.1.3 is well understood.

Actually, electrons are not excited to the conduction band from all the As atoms that Si is doped with. The probability of a neutral As, or the probability that an electron is not excited from the donor level of an As atom, is expressed by the probability that the electron occupies the donor level. Here, we denote the total concentration of a donor as N_d, the concentration of neutral donor as N_d^0, i.e. one without electron excitation, and the concentration of donor ions as N_d^+. The concentration of donor ions corresponds to the number of donor-level electrons produced by excitation, as expressed by this equation

$$N_d = N_d^+ + N_d^0. \quad (8.1.6)$$

Equation (8.1.6) can be rewritten by substituting the Fermi Dirac distribution function, $f(E)$, to yield Eq. (8.1.7),

$$N_d^+ = N_d - N_d^0 = N_d \left(1 - \frac{N_d^0}{N_d} \right) = N_d \left(1 - \frac{1}{1 + \frac{1}{2} \exp \frac{E_d - E_F}{k_B T}} \right). \quad (8.1.7)$$

Here, 1/2 is used in the denominator on the right side of Eq. (8.1.7) because we must not consider two electrons in one orbital of a donor but only one electron. Such a parameter is called a 'degeneracy factor (g)', $1/2 = 1/g$. At room temperature $(E_d - E_F \gg k_B T)$, Eq. (8.1.7) is simplified as follows,

$$N_d = N_d^+. \quad (8.1.8)$$

This implies that all donor atoms are ionized at room temperature.

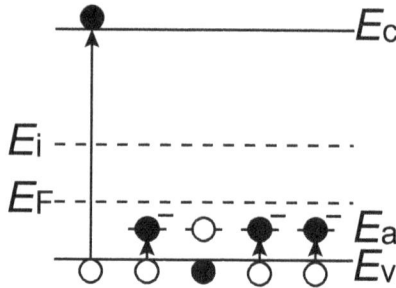

Fig. 8.1.4   Carrier generation in a p-type semiconductor.

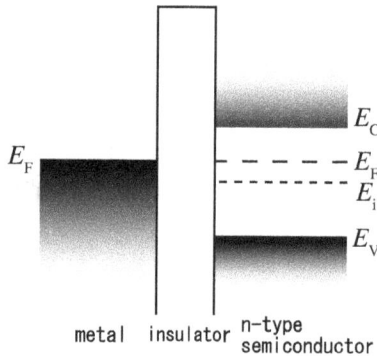

Fig. 8.1.5   Energy diagram of MOS structure. $V_G = 0$.

The energy diagram for Si doped with B atoms, which have one electron less than Si, is shown in Fig. 8.1.4. The E_a is the energy of the acceptor level in which that electron is accepted from the valence band. The Fermi level is different from that shown in Fig. 8.1.3. Thus, when Si is doped with B, holes are formed as follows:

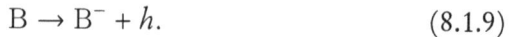

$$B \rightarrow B^- + h. \qquad (8.1.9)$$

A semiconductor in which holes are the majority carrier is called a "p-type semiconductor". It is important to note that the total charge is zero in both n- and p-type semiconductors.

We will consider FET operation using an n-type semiconductor. Figure 8.1.5 shows the energy diagram of an n-type semiconductor connected to a gate dielectric and gate electrode. This structure is called a "metal–oxide–semiconductor (MOS)", where metal, oxide and semiconductor refer to the gate electrode, gate dielectric and n-type semiconductor, respectively. Initially, no bias voltage is applied to the gate electrode, i.e. $V_G = 0$.

When $V_G > 0$, the Fermi level of the metal electrode decreases, and all energy levels, E_c, E_v and E_i, are bent downward, as shown in Fig. 8.1.6. Here, the Fermi level of the semiconductor does not change because there is no current flow from the semiconductor to the gate electrode. The downward band-bending at $V_G > 0$ must attract electrons to the interface between the semiconductor and the gate dielectric, producing an electron-rich region in comparison with the bulk of the semiconductor. This makes the concentration of electrons, which are the majority carrier in an n-type semiconductor, higher near the interface. The electron-enriched region is represented by the symbol "n". This situation is called electron accumulation. However, the MOSFET does not operate due to the accumulation. This is explained below.

As described later, the source/drain electrodes are prepared by generating carrier-enriched areas in a Si MOSFET. If electron-enriched areas (expressed as "n^+") are generated in the source/drain electrodes, the carrier type of the source–channel–drain will become n^+-n-n^+, resulting in the flow of electric current when $V_G > 0$. On the other hand, if hole-enriched areas (expressed as "p^+") are generated in the electrodes, the carrier type of the source–channel–drain will become p^+-n-p^+, resulting in no electric current flow when $V_G > 0$. Here, it should be noted that the channel does not change to p when $V_G > 0$ (accumulation) and $V_G = 0$. To sum up, electric current flows in n^+-n-n^+ condition or no electric current flow in p^+-n-p^+ condition, when $V_G \geq 0$. These results imply that accumulated channels cannot be used for FET operation. Thus, we do not discuss the operation of an FET based on an accumulated channel. As shown later, in practice only p^+ areas are generated for source/drain electrodes in n-type semiconductors.

When $V_G < 0$, upward band bending takes place, as seen in Fig. 8.1.6. Therefore, electrons (the majority carrier in n-type semiconductors) are repelled from the interface, i.e. the electrons move to the bulk, and carriers are depleted at the interface. Actually, only the donor ions, which

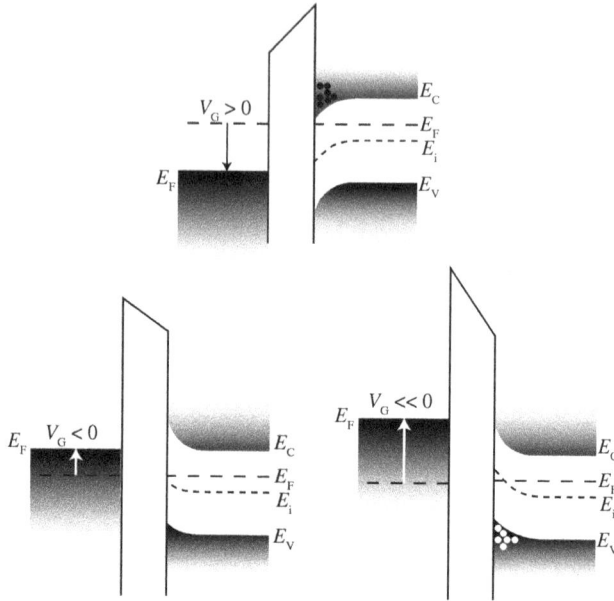

Fig. 8.1.6   Energy diagram of MOS structure. $V_G \neq 0$.

are positively charged, are located in the interface, because they are not mobile. This situation is called "depletion" because of the lack of carriers.

When $V_G \ll 0$, a rapid upward bending of energy levels occurs so that the E_i overcomes the E_F (see Fig. 8.1.6), which produces a hole-enriched region near the interface. The concentration of holes, which are the minority carrier of n-type semiconductors, becomes much higher than that in the bulk semiconductor. This situation is called 'inversion'. A Si MOSFET operates in the inversion state. The inversion region becomes the channel region of the FET, through which electric current flows between source and drain electrodes when a bias voltage is applied. To sum up, the minority carrier (holes in an n-type semiconductor) is responsible for charge transport in a Si MOSFET. Thus p-channel FET operation, in which hole movement constitutes the electric current, is achieved for an n-type semiconductor, whereas in n-channel operation for a p-type semiconductor it is moving electrons that constitute the current flow.

Here, we must discuss the carrier type in the source-channel-drain in a Si MOSFET. Now, we are considering an n-type semiconductor and an inversion layer (labelled "p" in an n-type semiconductor). If electron-enriched areas (labelled "n^+") are generated in the source/drain electrodes, the carrier type of the source–channel–drain will become n^+-p-n^+ by the formation of an inversion layer when $V_G \ll 0$, resulting in no current flow in the channel region. However, bulk conduction takes place, resulting in the flow of current, because this is an n^+-n-n^+ (n-type semiconductor). When $V_G \sim 0$, it will be n^+-n-n^+ in the entirety of the semiconductor, resulting in electric current flow. This implies that electric current flows when $V_G \leq 0$, i.e. FET operation is not expected.

On the other hand, if hole-enriched areas (expressed "p^+") are generated in the electrodes, the carrier type of the source–channel–drain will become p^+-p-p^+ when $V_G \ll 0$, resulting in electric current flow (switched-on state). The bulk current does not flow under the condition of p^+-n-p^+. When $V_G \sim 0$, it will be p^+-n-p^+ throughout the semiconductor, yielding the switched-off state. This shows how the on-off switching of electric current by varying V_G produces a normally-off FET. Thus, only p^+ electrodes can provide FET characteristics through the formation of an inversion layer in an n-type semiconductor. The same analysis can also be applied to a Si MOSFET with a p-type Si semiconductor, in which n^+ electrodes must be formed for source/drain electrodes.

As described later, various metals, with different work functions, are placed on the active layer to make source/drain electrodes in organic FETs. The gate electrode is arranged parallel to the active layer across the gate dielectric, as seen in Fig. 8.1.7, which shows an FET as a three-terminal device, consisting of source, drain and gate electrodes.

A Schottky barrier is generally formed between the active layer and the source/drain electrodes when the E_F of the metal electrode is lower (higher) than the Fermi level of the active layer in the case of n-channel operation (p-channel operation). Therefore, the drain current, I_D, measured in two-terminal measurement mode includes the contact resistance. The I_D can be measured by connecting an electrometer between the power supply and drain electrode. To remove contact resistance, another set of electrodes must be attached to detect the voltage drop in the channel region for what is called "four-terminal

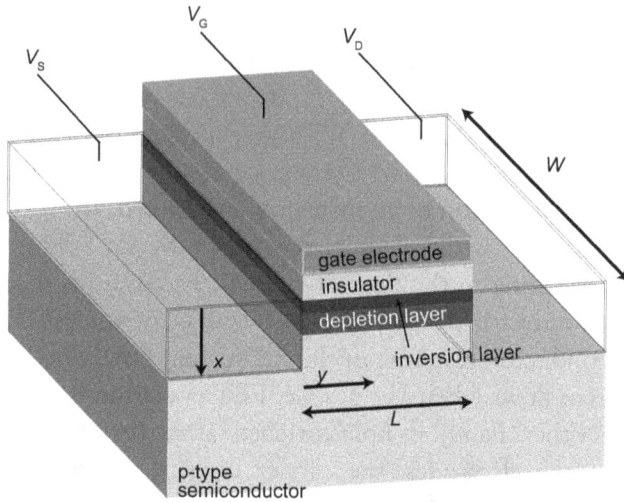

Fig. 8.1.7   Structure of Si MOSFET.

measurement". Namely, two electrodes for flowing electric current refer to source and drain electrodes, and another two electrodes are used for the detection of voltage drop. For ease of understanding, FET operation in two-terminal measurement mode will be discussed in this chapter.

We will consider a p-type semiconductor and n-channel operation. The source — drain voltage, V_{DS}, is expressed as follows because the source electrode is grounded ($V_S = 0$).

$$V_{DS} = V_D - V_S = V_D - 0 = V_D. \qquad (8.1.10)$$

V_D and V_S refer to drain voltage and source voltage, respectively. When the I_D is measured at the fixed V_{DS}, a very small I_D should be detected because of large resistance in the semiconductor. When applying positive V_G, the I_D should increase when the V_G exceeds the voltage (threshold voltage: V_{TH}) above which an inversion layer is formed. In the inversion layer, the minority carrier (electrons in p-type semiconductors) is enriched at the interface (channel region) near the gate dielectric. The resistance rapidly lowers in the channel region. The I_D which flows in the inversion layer (channel region) is expressed as follows:

$$I_D dy = \mu W C_O [V_G - V_{TH} - V(y)] dV, \qquad (8.1.11)$$

where y and W refer to distance from the edge of the source electrode and the width of the channel region (Fig. 8.1.7). C_O is the capacitance per area of the gate dielectric. By considering a constant I_D between source and drain electrodes, Eq. (8.1.11) is integrated over the range of $y = 0$ to L; L is the distance between source and drain electrodes, called the "channel length" (Fig. 8.1.7):

$$I_D \int_0^L dy = \int_0^{V_D} \mu W C_O [V_G - V_{TH} - V(y)] dV. \qquad (8.1.12)$$

As a consequence, the equation expressing I_D is obtained as follows:

$$I_D L = \mu W C_O \left[(V_G - V_{TH}) V_D - \frac{1}{2} V_D^2 \right]. \qquad (8.1.13)$$

When the V_D is small, the second term, $\frac{1}{2} V_D^2$, of Eq. (8.1.13) is ignored, and equation (8.1.14) is obtained.

$$I_D = \frac{\mu W C_O}{L} (V_G - V_{TH}) V_D. \qquad (8.1.14)$$

From this equation, the $I_D - V_G$ plot is drawn as shown in Fig. 8.1.8, when the V_D is small. This plot is called the "transfer characteristic". To obtain the μ value, the slope of the experimental $I_D - V_G$ plot must be evaluated.

The V_G value providing $I_D = 0$ becomes V_{TH}. Thus, the FET characteristic at V_D ($< V_G - V_{TH}$) corresponds to the linear regime, i.e. the FET characteristic in the linear regime is described by Eq. (8.1.14).

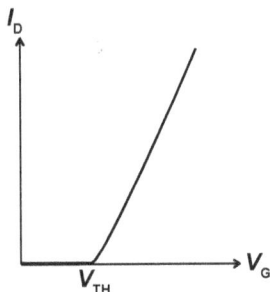

Fig. 8.1.8   Transfer plot in the linear regime. $V_D < V_G - V_{TH}$.

In the case of a large V_D, if the V_D satisfies this equation,

$$V_D > V_G - V_{TH},\qquad(8.1.15)$$

the following equation holds.

$$0 > (V_G - V_{TH}) - V_D.\qquad(8.1.16)$$

Therefore, an inversion layer cannot be formed near the drain electrode because of the negative effective gate voltage. Even with a large V_G, no inversion layer is generated when Eq. (8.1.16) holds. Thus, the channel between source and drain electrodes is not completely formed, an effect called "pinch-off". Namely, the I_D increases monotonically at V_D, satisfying Eq. (8.1.17)

$$V_D < V_G - V_{TH}.\qquad(8.1.17)$$

However, when the V_D increases and invalidates relation (8.1.17), the channel region does not expand, and the actual V_D follows equation (8.1.18).

$$V_D = V_G - V_{TH}.\qquad(8.1.18)$$

Thus, a V_D exceeding the value of $V_G - V_{TH}$ does not contribute to the formation of a channel region, and it is used as the bias voltage for getting over the depletion region formed near the drain electrode at pinch-off. Therefore, the V_D must become $V_G - V_{TH}$, and Equation (8.1.14) is rewritten as

$$I_D = \frac{\mu W C_O}{2L}(V_G - V_{TH})^2.\qquad(8.1.19)$$

At $V_D(> V_G - V_{TH})$, the FET characteristic must be plotted as $I_D^{1/2}$ *versus* V_G. The slope of the $I_D^{1/2} - V_G$ plot leads to the μ value from Eq. (8.1.19). The FET characteristic at $V_D(> V_G - V_{TH})$ corresponds to the saturation regime, i.e. the FET characteristic at saturation is described by equation (8.1.19). Figure 8.1.9 shows a typical $I_D^{1/2} - V_G$ plot.

The $I_D - V_D$ plots were plotted at different V_G values, and are called "output characteristics". As seen in Fig. 8.1.10, the I_D increases linearly against V_D at V_D ($< V_G - V_{TH}$), thus corresponding to the linear regime. On the other hand, the I_D becomes constant at V_D ($> V_G - V_{TH}$), which

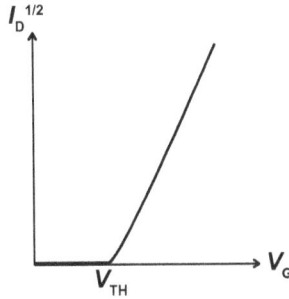

Fig. 8.1.9   Transfer plot in the saturation regime. $V_D > V_G - V_{TH}$.

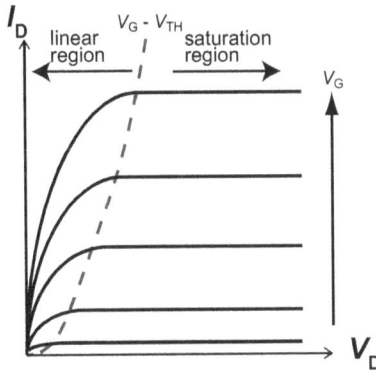

Fig. 8.1.10   Typical output properties.

corresponds to the saturation regime. As described previously, a channel region is not formed near the drain electrode in the saturation regime.

In the linear regime ($V_D < V_G - V_{TH}$), the I_D should increase linearly against V_D, but a concave $I_D - V_D$ plot is often observed because of the large contact resistance between source/drain electrodes and the active layer when the $I_D - V_D$ plot is measured in two-terminal measurement mode. Such a concave curve is frequently observed in a single-crystal FET, making it difficult to evaluate FET parameters. To avoid the appearance of a concave $I_D - V_D$ plot, we must reduce the contact resistance and use four-terminal measurement mode for the evaluation of FET properties. The reduction of contact resistance will be described in detail later.

We discussed n-channel FET characteristics using a p-type semiconductor, but p-channel FET characteristics using an n-type semiconductor can also be analysed using the same equation. Actually, because V_G, V_D, and I_D are negative, their absolute values must be used for analysis. Furthermore, the source electrode must be grounded ($V_S = 0$) in any operation mode, and the drain electrode should be connected to positive voltage for n-channel operation, while the drain electrode should be connected to negative voltage for p-channel operation.

8.1.2. Operation Properties of Organic FETs

Transport properties of an organic FET can be evaluated based on the analogy with a Si MOSFET. However, an organic FET has some operation mechanisms that differ from those of a Si MOSFET. An intrinsic semiconductor is not used for the active layer of a Si MOSFET, but an extrinsic semiconductor doped with an impurity is used. However, impurity doping is not used in the active layer of an organic FET, i.e. an intrinsic organic semiconductor is used for the active layer. Nevertheless, the purity of the organic material is much lower than that of an intrinsic Si semiconductor. For instance, the purity of non-doped Si is as high as 99.9999999% (nine 9's), but the purity of an organic crystal is at most 99.99%. Therefore, a trace of impurity must be present even if no doping is performed on organic materials (whether crystals or thin films). Furthermore, an organic FET is ambipolar, and can operate in both p-channel and n-channel modes by changing the electrode metal, contrary to the behaviour of a Si MOSFET, which operates as either p-channel or n-channel (unipolar). These characteristics found in organic FETs are different from those in a Si MOSFET, where the minority carrier plays a role in electric transport.

The thin-film FET that uses pentacene, which is a π-extended organic molecule, operates in p-channel by using Au for source/drain electrodes, and the μ reaches ~1 cm^2 V^{-1} s^{-1}. However, the thin-film FET using pentacene can also operate in n-channel mode by using Ca for source/drain electrodes, though the μ is as low as 10^{-3} cm^2 V^{-1} s^{-1}.

We must examine these phenomena. As seen in Fig. 8.1.11, the E_F values of pentacene and most π-extended organic compounds are not located at the centre position between E_c and E_v but close to the top

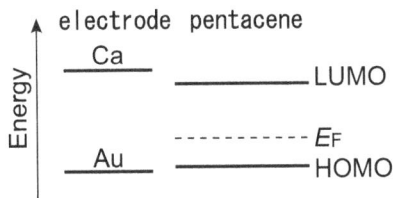

Fig. 8.1.11   Energy diagram of pentacene and metal electrodes.

of E_v; E_F is close to the HOMO level in Fig. 8.1.11. Therefore, a large V_G may be necessary to reach E_c because many trap states are located between E_F and E_c, which must be filled by electrons. On the other hand, the $|V_G|$ required to fill the trap states with holes should be small because of the small energy difference between E_F and E_v. This explains the easy observation of p-channel operation in a pentacene FET.

Moreover, we can stress that the energy barrier between the active layer and the source/drain electrodes should be so small that a hole can easily be injected from the source electrode and expelled to the drain electrode. This is because the energy difference between the E_F of Au and E_v is very small. In the case of pentacene, an energy barrier may not be formed because the HOMO level is located at a slightly higher energy than the E_F of Au, i.e. Ohmic contact between Au and the active layer is realized. These results imply that a hole can easily reach the valence band when driven by a small $|V_G|$, and that holes are continuously injected without any disturbance, guaranteeing stable FET operation.

When source/drain electrodes are formed by Ca metal, the energy difference between the E_F of Ca and its E_c becomes small, leading to a small energy barrier between electrodes and active layer. Consequently, the stable/easy injection and emission of electrons are guaranteed. Actually, Ohmic contact may be formed in the case of pentacene and Ca because of the higher energy of E_F of Ca than E_c, as seen in Fig. 8.1.11. This is the origin of n-channel operation in a pentacene thin-film FET with Ca source/drain electrodes. However, an imbalance between p-channel and n-channel operation is observed in pentacene thin-film FETs, owing to the fact that the E_F of pentacene is close to the E_v, i.e. pentacene is a p-type semiconductor. In addition, the concept of an inversion layer is not required to understand the FET properties of an organic FET. The fact that the effectiveness of injection (emission) of carriers from (to) an electrode changes depending on the work function of the electrode metal may be explained by the fact that the E_F value of organic materials is constant even if the metal electrode is varied. That is, the energy barrier between metal electrodes and the active layer changes when the metal electrodes are changed. Such an effect is not observed with Si. This can be explained with reference to the slope parameter S_x which is expressed as follows:

$$S_x = \frac{d\phi_B}{dX_m},$$
(8.1.20)

where ϕ_B and X_m refer to the energy barrier and electronegativity, respectively. S_x is an indication of how much the ϕ_B changes when changing metals. $S_x = 0$ for Si, and $S_x = 1$ for GaS.[17] $S_x = 0$ means 'Bardeen limit', while $S_x = 1$ means 'Mott-Schottky limit'. The organic compounds provide S_x values categorized as the Mott-Schottky limit. Therefore, the energy barrier in an organic FET is easily changed by changing the metal electrodes, which provides ambipolar properties. The transport properties of an organic FET are summarized as follows:

(1) Organic materials are used for the active layers of organic FETs without any impurity doping, i.e., intrinsic organic semiconductors without any artificial doping are used in FET devices. The concept of an inversion layer is not part of the mechanism of FET operation, which is different from a Si MOSFET. The channel region can be freely formed by either electrons or holes, depending on the electronic states of the active layer and source/drain electrodes. Therefore, ambipolar FET properties can be produced in organic FETs by changing the metal of the electrodes.

(2) Because of the low purity of organic thin films and crystals, the actual E_F is not located at the centre position between E_c and E_v, but is close to either E_c or E_v. This determines the operating polarity in organic FETs. In other words, either p-channel or n–channel operation is observed in an organic FET. Even if both operate, the

operation property is unbalanced rather than symmetric. This relates to the V_G value required to fill the trap states.

(3) After the channel has formed through carrier accumulation by applying V_G, it is necessary to lower the energy barrier between the source/drain electrodes and the active layer, for the stable injection of carriers from the source electrode, as well as for stable emission to the drain electrode. Therefore, the work function of metal electrodes must be controlled by the use of suitable metals. This way, the energy barrier can be suitably modified by varying the work function, because the S_x values of organic materials lie on the Mott-Schottky limit.

8.1.3. The Future Development of High-performance Organic FETs

Recently, some organic thin-film and single-crystal FETs with very high μ values have been reported, and the previous upper limit of ~1 cm^2 V^{-1} s^{-1} for the μ value in an organic FET was exceeded in the intense studies of the past decade. In this chapter, some recent materials of interest for high-performance organic FETs will be introduced. Dioctylbenzothenobenzothiophene (C8-BTBT) was reported by Takimiya *et al.*[18] This molecule's structure is shown in Fig. 8.1.1. C8-BTBT is a π-extended organic material containing thiophene rings in the molecular framework. The C8-BTBT thin-film FET showed a μ as high as 9.1 cm^2 V^{-1} s^{-1} when a thin film was formed by a solution process.[18] Furthermore, a bis(benzothieno)naphthalene (BBTN) thin-film FET exhibited a μ of 15.6 cm^2 V^{-1} s^{-1}, with a thin film made by thermal deposition. The molecular structure of BBTN is similar to that of C8-BTBT.[19] Recently, Bao *et al.* recorded 43 cm^2 V^{-1} s^{-1} in an FET using BBTN with different alkyl chains, formed by a solution process.[20] The formation process is a significant factor in making thin films of BTBT and related materials.

Our group has fabricated FET devices using phenacene molecules, which have an extended, repeating W-shaped structure of fused benzene rings, as seen in Fig. 8.1.1.[21-25] The phenacene molecules are named "picene" for 5 benzene rings, "[6]phenacene" for 6 benzene rings,

"[7]phenacene" for 7 benzene rings, "[8]phenacene" for 8 benzene rings, and "[9]phenacene" for 9 benzene rings. We have already synthesized phenacene molecules with 5–9 benzene rings. Very recently, the syntheses of [10]phenacene (10 benzene rings) and [11]phenacene (11 benzene rings) were achieved. Furthermore, their thin-film and single-crystal FETs were investigated using various gate dielectrics. These molecules are chemically stable because of their large HOMO-LUMO gap and deep HOMO level.[15] The large transfer integral in the *ab* plane predicted from quantum chemistry calculation suggested a large electric transport parallel to the *ab* plane.

Actually, the *ab* planes are stacked parallel to the gate dielectric when the thin film is made by thermal deposition, which is suitable for the fabrication of an FET device.[15,21-25] For phenacene, thin plate-type single crystals are generated, and the extended plate corresponds to the *ab* plane, which also leads to easy fabrication of FET devices.[15-29] The maximum μ value recorded for a phenacene thin-film FET is 21 cm^2 V^{-1} s^{-1} for a 3,10-ditetradecylpicene picene-$(C_{14}H_{29})_2$ thin-film FET.[30] The molecular structure of picene-$(C_{14}H_{29})_2$ is shown in Fig. 8.1.1.

Such a high μ value is probably obtained because of the presence of an extended π-electron system and suitable alkyl chains, which must produce a molecular packing suitable for conduction. On the other hand, the highest μ value among phenacene single-crystal FETs was 18 cm^2 V^{-1} s^{-1}, which was recorded in a [9]phenacene single-crystal FET.[29] The μ values for phenacene FETs are obtained from the FET characteristics measured in two-terminal measurement mode, which means that the μ is affected by contact resistance. In particular, the contact resistance in a phenacene single-crystal FET is so large that a 3-nm thick 2,3,5,6-tetrafluoro-7,7,8,8-tetracyanoquinodimethane (F_4TCNQ) layer was inserted between the Au electrodes and the active layer. This insertion is very effective for breaking the Schottky barrier. It is indispensable for the measurement of FET characteristics in two-terminal measurement mode.[27-29] The μ value of a phenacene single-crystal FET tends to increase with an increase in number of benzene rings.[29] This is probably due to increasing π-π

interaction between molecules facilitated by the larger number of benzene rings. The extension of the benzene network must increase the π-π interaction.

Thus, some organic FETs have recently shown μ values greater than 20 cm^2 V^{-1} s^{-1}. This implies that the organic FET has reached a performance level sufficient for practical application in some areas, and the study of the fabrication of flexible organic FETs has advanced rapidly. However, the performance of n-channel organic FETs is relatively low in comparison with that of p-channel. This problem should be solved as soon as possible for the development of high-performance complementary MOS (CMOS) logic gate circuits with organic thin films. The development of n-channel FETs that operate stably in ordinary air is probably the most urgent study. Studies on the fabrication of n-channel organic FETs are fully reported in Ref. 31.

To realize a high-performance organic FET, we must (1) find suitable new organic materials, (2) advance key technologies to design the FET, and (3) improve the interfaces constituting the FET. Furthermore, it will be important to investigate the FET devices microscopically, using various probing methods. The project of three-dimensional (3D) active-site science aims to provide a precise understanding the structure and electronic states of the special sites that govern FET characteristics. For instance, to investigate the surface structure of a crystal is of vital importance because the channel is formed at the interface between the active layer (crystal) and the gate dielectric. This is the core in the project of 3D active-site science. For this purpose, we must use new techniques such as X-ray crystal truncation rod (CTR) holography. The X-ray CTR holography of tetracene crystal showed a difference in crystal structure between bulk and surface, demonstrating that this technique is very useful when exploring the relationship between FET operation and the structure of the channel region.[32] Furthermore, probing the surface structure of a crystal under the operation of FET must be pursued, i.e. *in-situ* analysis of FET devices must also be a future research subject. To sum up, the project of 3D active-site science addresses the clarification of the structure and electronic states by microscopically targeting the sites that determine FET performance, and feeding the new information back into device fabrication. This project must surely advance the research on organic FET devices.

Finally, some textbooks (Refs. 33–35) are introduced, so that readers can accurately study semiconductor physics and semiconductor devices.

The authors greatly appreciate the assistance of Ms. Shino Hamao, Mr. Yuma Shimo, and Mr. Takahiro Mikami with manuscript preparation.

References

[1] F. Ebisawa, T. Kurokawa, and S. Nara, *J. Appl. Phys.* **54**, 3255 (1983).

[2] R. C. Haddon, *et al.*, *Appl. Phys. Lett.* **67**, 121 (1995).

[3] Y.-Y. Lin, *et al.*, *IEEE Electron. Device Lett.* **18**, 606 (1997).

[4] K. Ueno, *et al.*, *Nat. Mater.* **7**, 855 (2008).

[5] J. T. Ye, *et al.*, *Nat. Mater.* **9**, 125 (2010).

[6] V. Podzorov, V. M. Pudalov, and M. E. Gershenson, *Appl. Phys. Lett.* **82**, 1739 (2003).

[7] J. Takeya, *et al.*, *Jpn. J. Appl. Phys.* **44**, L1393 (2005).

[8] V. Podzorov, *et al.*, *Phys. Rev. Lett.* **95**, 226601 (2005).

[9] S. I. Machida, *et al.*, *Phys. Rev. Lett.* **104**, 156401 (2010).

[10] Q. Xin, *et al.*, *Phys. Rev. Lett.* **108**, 226401 (2012).

[11] D. C. Hoesterey and G. M. Letson, *J. Phys. Chem. Solids* **24**, 1609 (1963).

[12] D. Knipp, R. A. Street, and A. R. Volkel, *Appl. Phys. Lett.* **82**, 3907 (2003).

[13] N. Kawasaki, *et al.*, *Appl. Phys. Lett.* **94**, 043310 (2009).

[14] T. Sakanoue and H. Sirringhaus, *Nat. Mater.* **99**, 736 (2010).

[15] Y. Kubozono, *et al.*, *Eur. J. Inorg. Chem.* **24**, 3806 (2014).

[16] T. Yasuda, *et al.*, *Appl. Phys. Lett.* **85**, 2098 (2004).

[17] W. Mönch, "Electric properties of semiconductor interfaces" (Springer, 2004).

[18] C. Liu, *et al.*, *Adv. Mater.* **23**, 523 (2011).

[19] N. Kurihara, *et al.*, *Jpn. J. Appl. Phys.* **52**, 05DC11 (2013).

[20] Y. Yuan, *et al.*, *Nature Commun.* **5**, 3005 (2014).

[21] H. Okamoto, *et al.*, *J. Am. Chem. Soc.* **130**, 10470 (2008).

[22] N. Kawasaki, *et al.*, *Appl. Phys. Lett.* **96**, 113305 (2010).

[23] R. Eguchi, *et al.*, *Phys. Chem. Chem. Phys.* **15**, 20611 (2013).

[24] Y. Sugawara, *et al.*, *Appl. Phys. Lett.* **98**, 013303 (2011).

[25] H. Okamoto, *et al.*, *Sci. Rep.* **4**, 5330 (2014).

[26] N. Kawai, *et al.*, *J. Phys. Chem. C* **116**, 7893 (2012).

[27] X. He, *et al.*, *J. Phys. Chem. C* **118**, 5284 (2014).

[28] Y. Shimo, *et al.*, *J. Mater. Chem. C* **3**, 7370 (2015).

[29] Y. Shimo, *et al.*, *Sci. Rep.* **6**, 21008 (2016).

[30] H. Okamoto, *et al.*, *Sci. Rep.* **4**, 5330 (2014).

[31] A. Facchetti, *Materials Today* **10**, 28 (2007).

[32] H. Morisaki, *et al.*, *Nature Commun.* **5**, 5400 (2014).

[33] S. M. Sze, *Semiconductor Devices, Physics and Technology* (John Wiley & Sons, Inc., 2001).

[34] J.-P. Colinge and C. A. Colinge, *Physics of Semiconductor Devices* (Kluwer Academic Publishers, 2002).

[35] S. S. Li, *Semiconductor Physical Electronics* (Springer, 2006).

Subchapter 8.2

ORGANIC DEVICE AND BIOTECHNOLOGY MATERIAL

Hiroshi Sekiguchi

Japan Synchrotron Radiation Research Institute

Yuji C. Sasaki

The University of Tokyo

8.2.1. Application for biological materials

New conceptual method is firstly attempted to an inorganic substances having a relatively stable structure, and thereafter it is applied to organic substances and biological materials, and it advances while proving the effectiveness of the measurement method. For example, well-known X-ray structure determination method starts from the structure determination of Zinc sulfide crystal by M. F. von Laue in 1912, after the X-ray discovery by Röntgen in 1895. In the 1950's, X-ray diffraction techniques have also been widely used in structural biology to determine the structure of protein molecules. The determination of double helix structure of DNA, which is major component of heredity, had great contribution to modern biology (Fig. 8.2.1).[1] Findings associated with X-ray diffraction method is demonstrated by the inorganic materials, which would be subject of Nobel Prize in Physics, eventually, are used

197

Fig. 8.2.1   Structure of DNA.

in the structural analysis of organic molecules and biomolecules, which is the subject of the Nobel Prize in Chemistry.

In general, the biomaterials are not only nucleic acids and protein molecules as described above. Small peptides that serve as signal transmitter in biological systems and lipids that are major components of the lipid bilayer are also biomaterials. The analytical difficulty of these bio-related materials instead of inorganic materials are as follows, (1) Insulation properties, (2) destructibility to quantum beam, (3) functionality at single molecule level and (4) difficulty to be crystallized. Quantum beams of X-rays or neutron beams do not matter with respect to (1), however in the case of electron beam excitation or photoelectron detection, a conductive sample is required for precise measurements. It is necessary to acquire analytical information before and after the irradiation of the quantum beam in regard to (2), to confirm that there are no effects. (3) and (4) are serious problems when it is difficult to prepare a single crystal necessary for obtaining structure information.

In addition, since biomaterial usually consists of light elements, such as carbon, hydrogen, oxygen, etc, the scattering cross section for the quantum beam is small. Therefore, the larger size of crystal usually would be needed.

8.2.2. Molecular structure determination and functional analysis in biomaterials

As described above, we understand the inherent difficulty in determining the 3D structure of biological materials using quantum beams. Then, what is the significance of determining the molecular structure of the biomaterial system? One good example for describing importance of structural information is DNA whose structure was determined in 1950s. Currently we know that gene is a sequence of DNA that codes for a specific protein, however, it was unknown that gene is DNA or protein at the beginning of 1950s. R. Franklin who works at M. Wilkins' laboratory solved the structure of DNA, and J. Watson and F. Crick proposed the model of hereditary. DNA has the double helix structure as shown in Fig. 8.2.1. It consists of two polynucleotide strands in antiparallel directions and is linked via a hydrogen bond of a complementary base (A/T (adenine/thymine) and G/C (guanine/cytosine)) pair. Such complementarity is basis of replication of genetic information. In a more detailed study for the relationship between molecular structure and the function of the molecule it evaluates, how a point mutation of amino acid sequence changes the function of the protein.

It is true that structural information is important to understand the function of the protein, however, it was found that a number of intrinsically disordered proteins (IDP) lacks ordered stable three-dimensional structure. The methodology was focused on the determination of the stable structure of protein and such disordered structure's information was less attracted because of the technical difficulty. The disordered part of the protein makes it possible to bind multiple molecules, and to facilitate the turn-over the biological reaction, and it is known that 1/3 part of functional proteins has such disordered regions. Methods for evaluating the active site of these IDPs will be discussed in the future. Besides the molecular structure determination, the method for investigating the molecular fluctuation and structural instability of molecules would be important in the future.

8.2.3. Measurement of metal-containing protein molecules

In many functional biomolecule systems, metallic elements play an important role in the function expression of biomolecules. Indeed, about half of biomolecules *in vivo* use metals to express their functions.[2] For example, the PSII (photosystem II or water-plastoquinone oxidore-ductase) contains Mn_4Ca cluster, and the high efficient photosynthesis is realized through changing the valence number of Mn. Since such a metal element plays a role as an active site, researches on metal element peripheral sites have been actively conducted in the field of life science. Most recently, the more accurate and damage-free structural determination was carried out using XFEL (X-ray Free Electron Laser) sources.[3] From such a viewpoint, it is a very important approach to

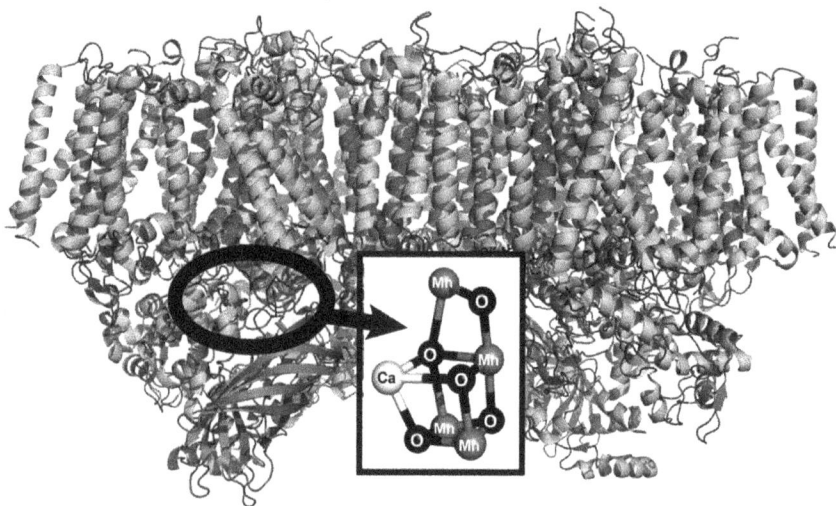

Fig. 8.2.2   Structure of PSII. Mn_4CaO_5 cluster, active site of PSII, was magnified.

elucidate the mechanism of function expression of biomolecules, by elucidating the local structure around the metal element by using the element selective method. In PSII analysis, one of important information is the change in the valence of Mn. Such change is difficult to determine directly by X-ray diffraction method using ordinary crystals. X-ray fluorescence holography has attracted attention. The X-ray fluorescence holography method is a relatively new structural analysis method that can visualize three-dimensional atomic images around specific elements. Unlike the X-ray diffraction structural analysis method which is powerful in periodic structure analysis and the X-ray absorption fine structure method which specializes in the analysis of the local structure up to 2–3 atoms ahead, up-to 2 nm local information could be reproduced by X-ray fluorescence holography method. For this reason, it can provide structure information unique to this method, such as valence-selective structure in a unique field called a local medium-range structure.[4]

8.2.4. Measurement method aiming at adaptation to general-purpose biomaterials

In order to adapt the measuring method, which originally developed for inorganic materials, to biomaterials, much of efforts are put for solving technical problems relating biomolecule. It would be better for us to recognize that the heavy element is not the main component for biomaterials, but the light element, such as hydrogen, and carbon, is the main component. One solution for that problem is to target sulfur S, which is a relatively heavy element, rather than carbon or hydrogen. The number of S atom is low in protein, however S–S disulfide bonds form a skeletal part of the structure of a protein molecule, and the determination of the arrangement of S–S bonds might equivalent to determining the protein molecular skeleton. In addition, the labelling method that could be realized by chelating with a metal ion at specific position in the protein molecule is one of options. Of course, when applying the labeling method, it is important to evaluate its biological functional effects for labelling. Non-labelling method will lead the future methodology when new phenomenon is discovered by the by the labelling method as described above.

References

[1] J. D. Watson and F. H. C. Crick, *Nature* **171**, 737 (1953).

[2] K. J. Waldron, *et al.*, *Nature* **460**, 823 (2009).

[3] M. Suga, *et al.*, *Nature* **517**, 99 (2015).

[4] K. Hayashi, *et al.*, *J. Phys. Condens. Matter* **24**, 093201 (2012).

INDEX